Foundations of Combinatorics

Foundations of Combinatorics

Desmond Ballard

STATES
ACADEMIC PRESS
www.statesacademicpress.com

Foundations of Combinatorics
Desmond Ballard
ISBN: 978-1-63989-215-0 (Hardback)

STATES
ACADEMIC PRESS

Published by States Academic Press,
109 South 5th Street,
Brooklyn, NY 11249, USA

Cataloging-in-Publication Data

Foundations of combinatorics / Desmond Ballard.
p. cm.
Includes bibliographical references and index.
ISBN 978-1-63989-215-0
1. Combinatorial analysis. 2. Algebra. 3. Mathematical analysis. I. Ballard, Desmond.
QA164 .C66 2022
511.6--dc23

For more information regarding States Academic Press and its products, please visit the publisher's website www.statesacademicpress.com

TABLE OF CONTENTS

PREFACE

This book is a culmination of my many years of practice in this field. I attribute the success of this book to my support group. I would like to thank my parents who have showered me with unconditional love and support and my peers and professors for their constant guidance.

The field of mathematics that primarily deals with counting is known as combinatorics. It is used as a means and an end in obtaining results and in studying properties of finite structures. It addresses the problems of enumeration, construction, existence and optimization of these structures. The discipline has many sub-fields. Some of these are analytic combinatorics, extremal combinatorics, enumerative combinatorics, probabilistic and algebraic combinatorics. Coding theory, discrete and computational geometry, combinatorics and physics, combinatorics and dynamical systems, and combinatorial optimization are some of the fields that are closely related to this discipline. The field finds its application in logical and statistical physics, pure mathematics, geometry, topology, computer science and evolutionary biology. The book studies, analyzes and upholds the pillars of combinatorics and its utmost significance in modern times. It also unfolds the innovative aspects of this area of study which will be crucial for the holistic understanding of the subject matter. This book, with its detailed analyzes and data, will prove immensely beneficial to professionals and students involved in this area at various levels.

The details of chapters are provided below for a progressive learning:

Chapter - Introduction

The field of mathematics that is associated with counting in obtaining results and properties of finite structures is termed as combinatorics. Permutation and combination, combinatorial geometry, probabilistic method, etc. are studied under its domain. This is an introductory chapter which will briefly introduce about combinatorics.

Chapter - Theories in Combinatorics

There are various theories that fall under combinatorics. Some of them are matroid theory, graph theory, order theory, discrete theory, combinatorial design theory, etc. This chapter closely examines these theories of combinatorics to provide an extensive understanding of the subject.

Chapter - Theorems in Combinatorics

Dilworth's theorem, Mirsky's theorem, Baranyai's theorem, Corners theorem, Folkman's theorem, Szemeredi's theorem, Kirchoff's theorem, Wagner's theorem, Hall's matching theorem, etc. are some of the theorems that are used within combinatorics. This chapter has been carefully written to provide an easy understanding of these theorems of combinatorics.

Chapter - Enumerative Combinatorics

The area of combinatorics that is concerned with the number of ways in which certain patterns can be created is called enumerative combinatorics. Generating function, alternating sign matrix, exponential formula, lattice path, etc. are a few of its concepts. All the aspects related to enumerative combinatorics have been carefully written to provide an easy understanding of the subject.

Chapter - Additive Combinatorics

Additive combinatorics is a special case of combinatorics that only uses the operations of addition and subtraction. It further includes Ruzsa triangle inequality, Gowers norm, sum-free sequence, restricted sumset, etc. This chapter delves into the subject of additive combinatorics for a thorough understanding of it.

Chapter - Algebraic Combinatorics

The area of mathematics that applies methods of abstract algebra, group theory and representation theory, to problems of combinatorics is called algebra combinatorics. Bender-Knuth involution, h-vector, Stanley's reciprocity theorem, Eulerian poset, Buekenhout geometry, etc. are some of its aspects. This chapter discusses algebraic combinatorics in detail.

Chapter - Diverse Areas of Combinatorics

There are various areas that fall under the domain of combinatorics such as analytic combinatorics, infinitary combinatorics, arithmetic combinatorics, etc. This chapter has been carefully written to provide an easy understanding of these diverse areas of combinatorics.

Desmond Ballard

CHAPTER 1

Introduction

The field of mathematics that is associated with counting in obtaining results and properties of finite structures is termed as combinatorics. Permutation and combination, combinatorial geometry, probabilistic method, etc. are studied under its domain. This is an introductory chapter which will briefly introduce about combinatorics.

Combinatorics, also called combinatorial mathematics is the field of mathematics concerned with problems of selection, arrangement, and operation within a finite or discrete system. Included is the closely related area of combinatorial geometry.

One of the basic problems of combinatorics is to determine the number of possible configurations (e.g., graphs, designs, arrays) of a given type. Even when the rules specifying the configuration are relatively simple, enumeration may sometimes present formidable difficulties. The mathematician may have to be content with finding an approximate answer or at least a good lower and upper bound.

In mathematics, generally, an entity is said to "exist" if a mathematical example satisfies the abstract properties that define the entity. In this sense it may not be apparent that even a single configuration with certain specified properties exists. This situation gives rise to problems of existence and construction. There is again an important class of theorems that guarantee the existence of certain choices under appropriate hypotheses. Besides their intrinsic interest, these theorems may be used as existence theorems in various combinatorial problems.

Finally, there are problems of optimization. As an example, a function f, the economic function, assigns the numerical value $f(x)$ to any configuration x with certain specified properties. In this case the problem is to choose a configuration x_0 that minimizes $f(x)$ or makes it $\varepsilon =$minimal—that is, for any number $\varepsilon > 0, f(x_0)\ f(x)+\varepsilon$, for all configurations x,with the specified properties.

FACTORIAL, PERMUTATION AND COMBINATION

Combinatorics can help us count the number of orders in which something can happen. Consider the following example: In a classroom there are 3 pupils and 3 chairs standing in a row. In how many different orders can the pupils sit on these chairs? Let us list the possibilities – in this example the 3 different pupils are represented by 3 different colours of the chairs.

There are 6 different possible orders. The number of possible orders increases very quickly as the number of pupils increases. With 6 pupils there are 720 different possibilities and it becomes impractical to list all of them. Instead we want a simple formula that tells us how many orders there are for n people to sit on n chairs. Then we can simply substitute 3, 4 or any other number for n to get the right answer.

Suppose we have 4 chairs and we want to place four pupils on them. There are 4 pupils who could sit on the first chair. Then there are 3 pupils who could sit on the second chair. There are 2 choices for the third chair, and only one choice for the final chair. In total, there are possibilities. To simplify notation, mathematicians use a "!" called factorial. For example, 5! ("five factorial") is the same as $5\times4\times3\times2\times1$. Above we have just shown that there are n! possibilities to order n objects.

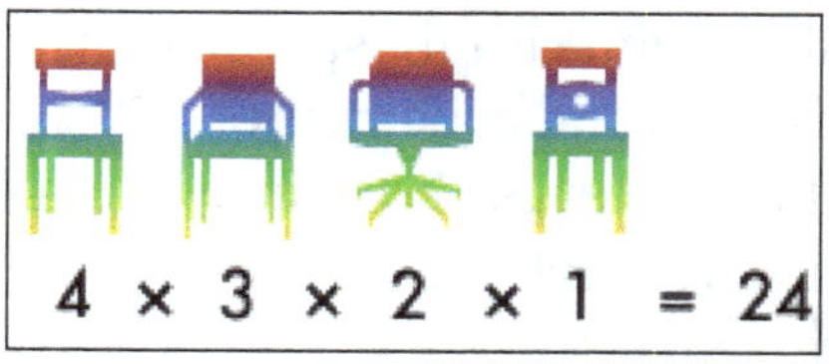

Permutations

The method above required us to have the same number of pupils as chairs to sit on. But what happens if there are not enough chairs?

How many different possibilities are there for any 2 of 3 pupils to sit on 2 chairs? Note that 1 will be left standing, which we don't have to include when listing the possibilities. Let us start again by listing all possibilities:

To find a simple formula like the one above, we can think about it in a very similar way. undefined In total there are possibilities. Again we should think about generalising this. We start like we would with factorials, but we stop before we reach 1. In fact we stop as soon as we reach the number students without chair. When placing 7 students on 3 chairs there are possibilities, since the $4\times3\times2\times1$ will cancel each other.

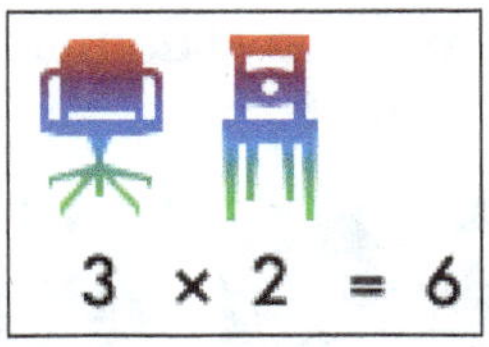

$$7\times6\times5=\frac{7\times6\times5\times4\times3\times2\times1}{4\times3\times2\times1}=\frac{7!}{4!}=\frac{7!}{(7-3)!}$$

Again there is a simpler notation for this: 7P3. If we want to place n objects in m positions there are possibilities.

$$nPm=\frac{n!}{(n-m)!}$$

The P stands for "permutations", since we are counting the number of permutations (orders) of objects. If m and n are the same, as they were in the problem at the beginning of this article, we have,

$$nPn=\frac{n!}{(n-n)!}=\frac{n!}{0!}$$

To make sense of this we define $0!=1$. Now, $nPn=n!$ as we would expect from our solution to the first problem.

Combinations

Permutations are used when you select objects and care about their order – like the order of children on chairs. However in some problems you don't care about the order and just want to know how many ways there are to select a certain number of objects from a bigger set.

In a shop there are five different T-shirts you like, coloured red, blue, green, yellow and black. Unfortunately you only have enough money to buy three of them. How many ways are there to select three T-shirts from the five you like?

Here we don't care about the order (it doesn't matter if we buy black first and then red or red first and then black), only about the number of combinations of T-shirts. The possibilities are so there are 10 in total.

If we had calculated 5P3 = 60, we would have double-counted some possibilities, as the following table shows:

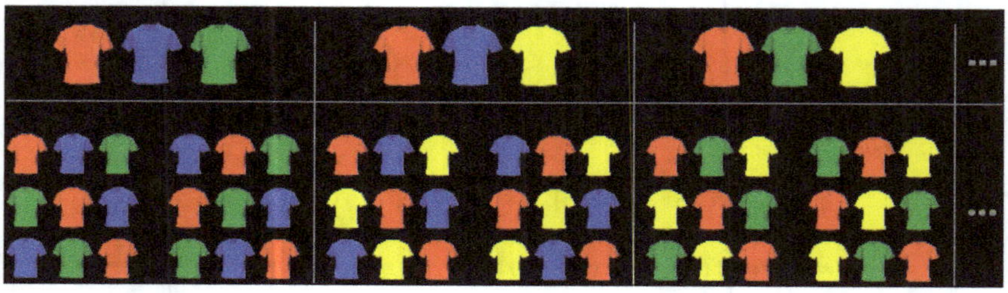

With permutations, we count every combination of three T-shirts 6 times, because there are $3! = 6$ ways to order the three T-shirts. To get the number of combinations from the number of permutations we simply need to divide by 6. We write,

$$5c3 = \frac{5P3}{3!} = \frac{60}{6} = 10.$$

Here the C stands for "combinations". In general, if we want to choose r objects from a total of n there are different combinations.

$$ncr = \frac{nPr}{r!} = \frac{n!}{r!(n-r)!}$$

Instead of nCr mathematicians often write $nCr = \left(\frac{n}{r}\right)$, like a fraction in brackets but without the line in between.

BASIC COMBINATORICS RULES

Suppose there are two sets A and B. The basic rules of combinatorics one must remember are:

Rule of Product

The product rule states that if there are X number of ways to choose one element from

A and Y number of ways to choose one element from B, then there will be $X \times Y$ number of ways to choose two elements, one from A and one from B.

Rule of Sum

The sum rule states that if there are X number of ways to choose one element from A and Y number of ways to choose one element from B, then there will be $X + Y$ number of ways to choose one element that can belong to either A or to B. These rules can be used for a finite collections of sets.

Permutations with Repetition

If we have N objects out of which N_1 objects are of type 1, N_2, objects are of type 2,... N_k objects are of type k, then number of ways of arrangement of these N objects are given by:

$$\frac{N!}{N_1!N_2!\ldots N_k!}$$

Combinations with Repetition

If we have N elements out of which we want to choose K elements and it is allowed to choose one element more than once, then number of ways are given by:

$$^{N+K-1}C_k = \frac{(N+K-1)!}{(K)!(N-1)!}$$

COMBINATORIAL OPTIMIZATION

Combinatorial Optimization is a branch of Mathematical Optimization with a vast number of applications. Be it the navigation system in your car, the software used to create timetables for high schools, or decision support systems in production and logistic environments, you can be almost certain that modern Combinatorial Optimization techniques are employed.

Combinatorial Optimization is concerned with finding an optimal or close to optimal solution among a finite collection of possibilities. The finite set of possible solutions is typically described through mathematical structures, like graphs, matroids or independence systems. The focus in Combinatorial Optimization lies on efficient algorithms which, more formally, means algorithms with a running time bounded by a polynomial in the input size. Therefore, two of the arguably most prominent questions in Combinatorial Optimization are:

- How quickly can one find a single (or all) optimal solutions of a given problem?

- When dealing with a problem where, due to complexity-theoretic reasons, it is unlikely that an optimal solution can be found efficiently: What is the best solution quality that an efficient algorithm can guarantee?

Over the last decades,Combinatorial Optimization has grown into a very mature field with strong links to various other disciplines like discrete mathematics (graph theory, combinatorics), computer science (data structures, complexity theory), probability theory, continuous optimization, and many application areas. Advances in modern Combinatorial Optimization thus often happen through clever combinations of ideas from several fields.

COMBINATORIAL GEOMETRY

The name combinatorial geometry, first used by Swiss mathematician Hugo Hadwiger, is not quite accurately descriptive of the nature of the subject. Combinatorial geometry does touch on those aspects of geometry that deal with arrangements, combinations, and enumerations of geometric objects; but it takes in much more. The field is so new that there has scarcely been time for it to acquire a well-defined position in the mathematical world. Rather it tends to overlap parts of topology (especially algebraic topology), number theory, analysis, and, of course, geometry. The subject concerns itself with relations among members of finite systems of geometric figures subject to various conditions and restrictions. More specifically, it includes problems of covering, packing, symmetry, extrema (maxima and minima), continuity, tangency, equalities, and inequalities, many of these with special emphasis on their application to the theory of convex bodies. A few of the fundamental problems of combinatorial geometry originated with Newton and Euler. The majority of the significant advances in the field, however, have been made since the 1940s.

Among those branches of mathematics that interest serious working mathematicians, combinatorial geometry is one of the few branches that can be presented on an intuitive basis, without recourse by the investigator to any advanced theoretical considerations or abstractions.

Yet the problems are far from trivial, and many remain unsolved. They can be handled only with the aid of the most careful and often delicate reasoning that displays the variety and vitality of geometric methods in a modern setting. A few of the answers are natural and are intuitively suggested by the questions. Many of the others, however, require proofs of unusual ingenuity and depth even in the two-dimensional case. Sometimes a plane solution may be readily extendible to higher dimensions, but sometimes just the opposite is true, and a three-dimensional or n-dimensional problem may be entirely different from its two-dimensional counterpart. Each new problem must be attacked individually. The continuing charm and challenge of the subject are at least in part due to the relative simplicity of the statements coupled with the elusive nature of their solutions.

Packing and Covering

It is easily seen that six equal circular disks may be placed around another disk of the same size so that the central one is touched by all the others but no two overlap and that it is not possible to place seven disks in such a way. In the analogous three-dimensional situation, around a given ball (solid sphere) it is possible to place 12 balls of equal size, all touching the first one but not overlapping it or each other. One such arrangement may be obtained by placing the 12 surrounding balls at the midpoints of edges of a suitable cube that encloses the central ball; each of the 12 balls then touches four other balls in addition to the central one. But if the 12 balls are centred at the 12 vertices of a suitable regular icosahedron surrounding the given ball, there is an appreciable amount of free space between each of the surrounding balls and its neighbours. (If the spheres have radius 1, the distances between the centres of the surrounding spheres are at least $2/\cos 18° = 2.1029 \cdots$.) It appears, therefore, that by judicious positioning it might be possible to have 13 equal non-overlapping spheres touch another of the same size. This dilemma between 12 and 13, one of the first nontrivial problems of combinatorial geometry, was the object of discussion between Isaac Newton and David Gregory in 1694. Newton believed 12 to be the correct number, but this claim was not proved until 1953. The analogous problem in four-dimensional space was solved in 2003, the answer being 24.

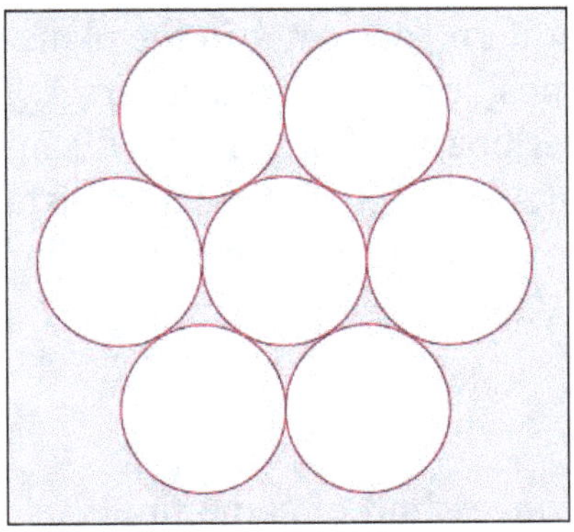

Packing of disks.

The problem of the 13 balls is a typical example of the branch of combinatorial geometry that deals with packings and coverings. In packing problems the aim is to place figures of a given shape or size without overlap as economically as possibly, either inside another given figure or subject to some other restriction.

Problems of packing and covering have been the objects of much study, and some striking conclusions have been obtained. For each plane convex set k, for example, it is possible to arrange nonoverlapping translates of K so as to cover at least two-thirds of the plane; if k is a triangle (and only in that case), no arrangement of nonoverlapping translates covers more than two-thirds of the plane. Another famous problem was Kepler's conjecture, which concerns the densest packing of spheres. If the spheres are packed in cannonball fashion—that is, in the way cannonballs are stacked to form a triangular pyramid, indefinitely extended—then they fill $\Pi / \sqrt{18}$, or about 0.74, of the space. In 1611 the German astronomer Johannes Kepler conjectured that this is the greatest density possible, but it was proved only in 1998 by the American mathematician Thomas Hales.

Covering of part of a plane with triangles.

Covering problems deal in an analogous manner with economical ways of placing given figures so as to cover (that is, contain in their union) another given figure. One famous covering problem, posed by the French mathematician Henri Lebesgue in 1914, is still unsolved: What is the size and shape of the universal cover of least area? Here a convex set C is called universal cover if for each set A in the plane such that diam A 1 it is possible to move C to a suitable position in which it covers A. The diameter diam A of a set A is defined as the least upper bound of the mutual distances of points of the set A. If A is a compact set, then diam A is simply the greatest distance between any two points of A. Thus, if A is an equilateral triangle of side 1, then diam $A = 1$; and if B is a cube of edge length 1, then diam $B = \sqrt{3}$.

Polytopes

A (convex) polytope is the convex hull of some finite set of points. Each polytope of dimensions d has as faces finitely many polytopes of dimensions 0 (vertices), 1 (edge), 2 (2-faces),..., d-1 (facets). Two-dimensional polytopes are usually called polygons, three-dimensional ones polyhedra. Two polytopes are said to be isomorphic, or of the same combinatorial type, provided there exists a one-to-one correspondence between their faces, such that two faces of the first polytope meet if and only if the corresponding faces of the second meet. The prism and the truncated pyramid of figure are isomorphic, the correspondence being indicated by the letters at the vertices. To classify the convex polygons by their combinatorial types, it is sufficient to determine the number of vertices υ; for each $u \geq 3$, all polygons with υ vertices (υ-gons) are of the same combinatorial type, while a υ-gon and a υ′-gon are not isomorphic if $u \neq u'$. Euler was the first to investigate in 1752 the analogous question concerning polyhedra. He found that υ – e + f = 2 for every convex polyhedron, where υ, e, and f are the numbers of vertices, edges, and faces of the polyhedron. Though this formula became one of the starting points of topology, Euler was not successful in his attempts to find a classification scheme for convex polytopes or to determine the number of different types for each υ. Despite efforts of many famous mathematicians since Euler, the problem is still open for polyhedra with more than 19 vertices. The numbers of different types with four, five, six, seven, or eight vertices are 1, 2, 7, 34, and 257, respectively. There are 2,606 different combinatorial types of convex polyhedra with nine vertices. The number of different types for 18 vertices is more than 107 trillion.

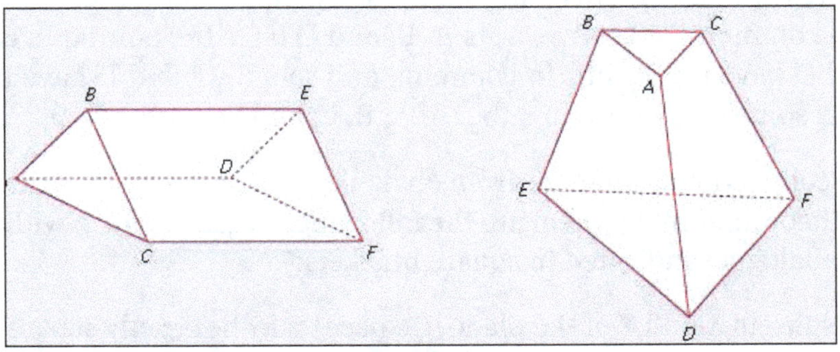

(Left) prism and (right) truncated pyramid.

The theory of convex polytopes has been successful in developments in other directions. The regular polytopes have been under investigation since 1880 in dimensions higher than three, together with extensions of Euler's relation to the higher dimensions. The interest in regular polyhedra and other special polyhedra goes back to ancient Greece, as indicated by the names Platonic solids and Archimedean solids.

Since 1950 there has been considerable interest, in part created by practical problems related to computer techniques such as linear programming, in questions of the following type: for polytopes of a given dimension d and having a given number υ of vertices, how large and how small can the number of facets be? Such problems have provided great impetus to the development of the theory. The U.S. mathematician Victor L. Klee solved the maximum problem in 1963 in most cases (that is, for all but a finite number of υ's for each d), but the remaining cases were disposed of only in 1970 by P. McMullen, in the United States, who used a completely new method.

Incidence Problems

If a finite set S of points in a plane has the property that each line determined by two points of S meets at least one other point of S, must all points of S be on one line? Sylvester never found a satisfactory solution to the problem, and the first (affirmative) solutions were published a half century later. Since then, Sylvester's problem has inspired many investigations and led to many other questions, both in the plane and in higher dimensions.

Helly's Theorem

Eduard Helly proved the following theorem, which has since found applications in many areas of geometry and analysis and has led to numerous generalizations, extensions and analogues known as Helly-type theorems. If $K_1, K_2, \cdots, K_n$ are convex sets in d-dimensional Euclidean space E^d, in which $n \geq d + 1$, and if for every choice of d + 1 of the sets Ki there exists a point that belongs to all the chosen sets, then there exists a point that belongs to all the sets $K_1, K_2, \cdots, K_n$. The theorem stated in two dimensions is easier to visualize and yet is not shorn of its strength: If every three of a set of n convex figures in the plane have a common point (not necessarily the same point for all trios), then all n figures have a point in common. If, for example, convex sets A, B, and C have the point p in common, and convex sets A, B, and D have the point q in common, and sets A, C, and D have the point r in common, and sets B, C, and D have the point s in common, then some point x is a member of A, B, C, and D.

Although the connection is often far from obvious, many consequences may be derived from Helly's theorem. Among them are the following, stated for d = 2 with some higher dimensional analogues indicated in square brackets:

A. Two finite subsets X and Y of the plane [d-space] may be strictly separated by a suitable straight line [hyperplane] if and only if, for every set Z consisting of at most 4 [d + 2] points taken from X ∪ Y, the points of X ∩ Z may be strictly separated from those of Y ∩ Z. (A line [hyperplane] L strictly separates X and Y if X is contained in one of the open half planes [half spaces] determined by L and if Y is contained in the other.)

B. Each compact convex set K in the plane [d-space] contains a point P with the following property: each chord of K that contains P is divided by P into a number of segments so the ratio of their lengths is at most 2d.

C. If G is an open subset of the plane [d-space] with finite area [d-dimensional content], then there exists a point P, such that each open half plane [half space] that contains P contains also at least 1/3 [1/(d + 1)] of the area [d-content] of G.

D. If $I_1, \cdots,$ In are segments parallel to the y-axis in a plane with a coordinate system (x, y), and if for every choice of three of the segments there exists a straight line intersecting each of the three segments, then there exists a straight line that intersects all the segments $I_1, \cdots,$ In.

Theorem D has generalizations in which kth degree polynomial curves y = akxk + · · · + $a_1x + a_0$ take the place of the straight lines and k + 2 replaces 3 in the assumptions. These are important in the theory of best approximation of functions by polynomials.

Methods of Combinatorial Geometry

Many other branches of combinatorial geometry are as important and interesting but rather than list them here it is more instructive to provide a few typical examples of frequently used methods of reasoning. Because the emphasis is on illustrating the methods rather than on obtaining the most general results, the examples will deal with problems in two and three dimensions.

Exhausting the Possibilities

Using the data available concerning the problem under investigation, it is often possible to obtain a list of all potential, a priori possible, solutions. The final step then consists in eliminating the possibilities that are not actual solutions or that duplicate previously found solutions. An example is the proof that there are only five regular convex polyhedra (the Platonic solids) and the determination of what these five are.

From the definition of regularity it is easy to deduce that all the faces of a Platonic solid must be congruent regular k-gons for a suitable k, and that all the vertices must belong to the same number j of k-gons. Because the sum of the face angles at a vertex of a convex polyhedron is less than 2π, and because each angle of the k-gon is $(k - 2)\pi/k$, it follows that $j(k - 2)\pi/k < 2\pi$, or $(j - 2)(k - 2) < 4$. Therefore, the only possibilities for the pair (j, k) are (3, 3), (3, 4), (3, 5), (4, 3), and (5, 3). It may be verified that each of these pairs actually corresponds to a Platonic solid, namely, to the tetrahedron, the cube, the dodecahedron, the octahedron, and the icosahedron, respectively. Very similar arguments may be used in the determination of Archimedean solids and in other instances.

The most serious drawback of the method is that in many instances the number of potential (and perhaps actual) solutions is so large as to render the method unfeasible. Therefore, sometimes the exact determination of these numbers by the method just discussed is out of the question, certainly if attempted by hand and probably even with the aid of a computer.

Use of Extremal Properties

In many cases the existence of a figure or an arrangement with certain desired properties may be established by considering a more general problem (or a completely different problem) and by showing that a solution of the general problem that is extremal in some sense provides also a solution to the original problem. Frequently there seems to be very little connection between the initial question and the extremal problem. As an illustration the following theorem will be proved: If K is a two-dimensional compact convex set with a centre of symmetry, there exists a parallelogram P containing K, such that the midpoints of the sides of P belong to K. The proof proceeds as follows: Of all the parallelograms that contain K, the one with least possible area is labeled P_0. The existence of such a P_0 is a consequence of the compactness of K and may be established by standard arguments. It is also easily seen that the centres of K and P_0 coincide. The interesting aspect of the situation is that P_0 may be taken as the P required for the theorem. In fact, if the midpoints A′ and A of a pair of sides of P_0 do not belong to K, it is possible to strictly separate them from K by parallel lines L′ and L that, together with the other pair of sides of P_0, determine a new parallelogram containing K but with area smaller than that of P_0. The above theorem and its proof generalize immediately to higher dimensions and lead to results that are important in functional analysis.

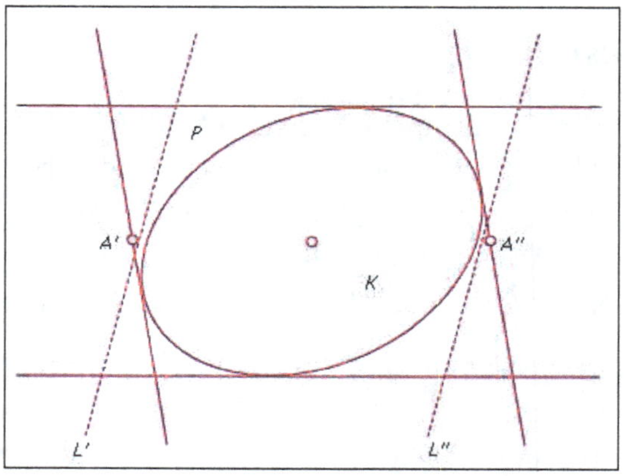

Example of theorem on extremal properties.

Sometimes this type of argument is used in reverse to establish the existence of certain objects by disproving the possibility of existence of some extremal figures. As an example the following solution of the problem of Sylvester. By a standard argument of projective geometry (duality), it is evident that Sylvester's problem is equivalent to the question: If through the point of intersection of any two of n coplanar lines, no two of which are parallel, there passes a third, are the n lines necessarily concurrent? To show that they must be concurrent, contradiction can be derived from the assumption that they are not concurrent. If L is one of the lines, then not all the intersection points lie on L. Among the intersection points not on L, there must be one nearest to L, which can be called A. Through A pass at least three lines, which meet L in points B, C, D, so that C is between B and D. Through C passes a line L* different from L and from the line through A. Since L* enters the triangle ABD, it intersects either the segment AB or the segment AD, yielding an intersection point nearer to L than the supposedly nearest intersection point A, thus providing the contradiction.

The difficulties in applying this method are caused in part by the absence of any systematic procedure for devising an extremal problem that leads to the solution of the original question.

Use of Transformations between Different Spaces and Applications of Helly's Theorem

The methods of proof in combinatorial geometry may be illustrated in one example—the proof of a theorem concerning parallel segments. Let the segment Ii have endpoints (x_i, y_i) and (x_i, y'_i), where y_i y'_i and i = 1, 2, · · ·, n. The case that two of the segments are on one line is easily disposed of; so it may be assumed that $x_1, x_2, \cdots, x_n$ are all different. With each straight line y = ax + b in the (x, y)-plane can be associated a point (a, b) in another plane, the (a, b)-plane. Now, for i = 1, 2, · · ·, n, the set consisting of all those points (a, b) for which the corresponding line y = ax + b in the (x, y) plane meets the segment Ii can be denoted by Ki. This condition means that $y_i\, ax_i + b\, y\, 'i$ so that each set K_i is convex. The existence of a line intersecting three of the segments I_i means that the corresponding sets K_i have a common point. Then Helly's theorem for the (a, b)-plane implies the existence of a point (a*, b*) common to all sets K_i. This in turn means that the line y = a*x + b* meets all the segments I_i, I_2, · · ·, In, and the proof of theorem D is complete.

In addition to the methods, many other techniques of proof are used in combinatorial geometry, ranging from simple mathematical induction to sophisticated decidability theorems of formal logic. The variety of methods available and the likelihood that there are many more not yet invented continue to stimulate research in this rapidly developing branch of mathematics.

PROBABILISTIC METHOD

Probabilistic method is a general methodology developed by Paul Erdös starting in the late 1940s. The idea of the method is that in order to prove the existence of an object one designs a non-deterministic algorithm that may or may not produce the desired object. Having the probability of failure less than one guarantees the existence of the object.

Let A_k, k = 1,..., m be subsets of a set Ω, each with n elements: $|A_k| = n$, k = 1,..., m. If $m < 2^{n-1}$, then there exists a bichromatic coloring of Ω such that no A_k is monochromatic.

Proof: Let F be a collection of n-sets (sets with exactly n elements), and assume that $|F| = m < 2^{n-1}$. Color Ω randomly with two colors, all colorings being equally likely. For A ∈ F let E_A be the event that A is monochromatic. Since there are two such colorings and |A| = n, probability $P(E_A)$ of the event E_A is given by:

$$P(E_A) = 2\times 2^{-n} = 2^{1-n}.$$

Since the events E_A are not necessarily disjoint,

$$P(\cup_{A\in F} E_A) < \sum_{A\in F} P(E_A) = m2^{1-n} < 1.$$

So the probability that at least one A ∈ F is monochromatic is less than 1. Thus there bound to be a bichromatic coloring with no monochromatic A's. For example, if |Ω| = m = 2n-1 then, for any bichromatic coloring of Ω, there is a monochromatic subset A, |A| = n. We may apply similar reasoning to the problem of awards in a basketball tournament.

If the results of the tournament are random, then the probability that there is no team that won against a selection of n teams is $(1 - 2^{-n})^{m-n}$. There are C(n, m) such selections. This leads a condition:

$$C(m, n)(1 - 2^{-n})^{m-n} < 1.$$

If this condition holds, there is a non-zero probability that, for every selection of n teams, there is a team that beats them all.

Lower Bounds on the Ramsey Number R(n, n)

Ramsey theory, roughly stated, is the study of how "order" grows in systems as their size increases. In the language of graph theory, the central result of Ramsey theory is the following:

Theorem (Ramsey, Erd¨os-Szekeres) Given a pair of integers s, t, there is an integer R(s, t) such that if n ≥ R(s, t), any 2-coloring of K_n's edges must yield a red Ks or a blue K_t.

A fairly simple recursive upper bound on R(s, t) is given by:

$$R(s,\ t) \le R(s, t-1) + R(s-1, t)$$

which gives us,

$$R(s,\ t) \le \binom{k+l-2}{k-1}$$

and thus, asymptotically, that,

$$R(s,\ s) \le 2^{2s} \cdot s^{-1/2}$$

A constructive lower bound on R(s, s), discovered by Nagy, is the following:

$$R(s,\ s) \ge \binom{s}{3}$$

(Explicitly, his construction goes as follows: take any set S, and turn the collection of all 3-element subsets of S into a graph by connecting subsets iff their intersection is odd).

Theorem: $R(s,\ s) > \lfloor 2^{s/2} \rfloor$.

Proof: Fix some value of n, and consider a random uniformly-chosen 2-coloring of Kn's edges: in other words, let us work in the probability space (Ω, P r) = (all 2-colorings of,

K_n's edges, $P\,r(\omega) = 1/2^{\binom{n}{2}}$.)

For some fixed set R of s vertices in V (K_n), let AR be the event that the induced subgraph on R is monochrome. Then, we have that,

$$\mathbb{P}(A_R) = 2.\left(2^{\binom{n}{2}-\binom{s}{2}}\right)/2^{\binom{n}{2}} = 2^{1-\binom{s}{2}}.$$

Thus, we have that the probability of at least one of the A_R's occuring is bounded by,

$$\mathbb{P}\left(\bigcup_{|R|=s} A_R\right) \leq \sum_{R\subset\Omega,|R|=s} \mathbb{P}(A_R) = \binom{n}{s} 2^{1-\binom{k}{2}}.$$

If we can show that $\binom{n}{s} 2^{1-\binom{s}{2}}$ is less that 1, then we know that with nonzero probability there will be some 2-coloring $\omega \in \Omega$ in which none of the A_R's occur. In other words, we know that there is a 2-coloring of K_n that avoids both a red and a blue K_s.

Solving,

$$\binom{n}{s} 2^{1-\binom{s}{2}} < \frac{n^s}{s!} \cdot 2^{1+(s/2)-(s^2/2)} = \frac{2^{1+s/2}}{s!} \cdot \frac{n^s}{2^{s^2/2}} < 1$$

Whenever, $n = \lfloor 2^{s/2} \rfloor, s \geq 3$.

Tournaments and the S_k Property

A tournament is simply an oriented K_n; in other words, it's a directed graph on n vertices where for every pair (i, j), there is either an edge from i to j or from j to i, but not both.

A tournament T is said to have property S_k if for any set of k vertices in the tournament, there is some vertex that has a directed edge to each of those k vertices.

One natural question to ask about the S_k property is the following:

- How small can a tournament be if it satisfies the S_k property, for some k?

We can calculate values of S_k for the first three values by hand:

- If k = 1, a tournament will need at least 3 vertices to satisfy S_k (take a directed 3-cycle).
- If k = 2, a tournament will need at least 5 vertices to satisfy S_k.
- If k = 3, a tournament will need at least 7 vertices to satisfy S_k (related to the Fano plane).

For k = 4, constructive methods have yet to find an exact answer; as well, constructive methods have been fairly bad at finding asymptotics for how these values grow. Probabilistic methods, however, give us the following useful bound:

Proposition: (Erdös) There are tournaments that satisfy property S_k on $O(k^2 2^k)$- many vertices.

Proof: Consider a "random" tournament: In other words, take some graph K_n, and for every edge (i, j) direct the edge $i \to j$ with probability 1/2 and from $j \to i$ with probability 1/2.

Fix a set S of k vertices and some vertex $v \notin S$. What is the probability that v has an edge to every element of S? Relatively simple: in this case, it's just $1/2^k$.

Consequently, the probability that v fails to have a directed edge to each member of S is $1 - 1/2^k$. For different vertices, these events are all independent; so we know in fact that,

$$\mathbb{P}(\text{for all } v \notin S, v \not\to S) = \left(1 - 1/2^k\right)^{n-k}.$$

There are $\binom{n}{k}$-many such possible sets S; so, by using a naive union upper bound, we have that,

$$\mathbb{P}\left(\exists S \text{ such that } \forall v \notin S, v \not\to S\right) \le \binom{n}{k} \cdot \left(1 - 1/2^k\right)^{n-k}$$

Thus, it suffices to force the right-hand side to be less than 1, as this means that there is at least one graph on which no such subsets S exist – i.e. that there is a graph that satisfies the S_k property.

So, using the approximation $\binom{n}{k} \cdot \left(1 - 1/2^k\right)^{n-k} \le \left(\frac{en}{k}\right)^k$, we calculate:

$$\left(e^{-1/2k}\right)^{n-k} < 1$$

$$\Leftrightarrow \left(\frac{en}{k}\right)^k < e^{(n-k)/2^k}$$

$$\Leftrightarrow k\left(1 + \log(n/k)\right) . 2^k + k < n$$

Motivated by the above, take $n > 2k \cdot k$; this allows us to make the upper bound,

$$k\left(1 + \log(n/k)\right).2^k + k < k\left(1 + \log\left(k2^k/k\right)\right) . 2^k + k$$

$$= 2^k \ . \ k^2 \ . \log(2) . \left(1 + \frac{1}{k\log(2)} + \frac{1}{k2^k \log(2)}\right)$$

$$= k^2 2^k \log(2) . \left(1 + O(1)\right);$$

so, if $n > k^2 2^k \log(2) \cdot (1 + O(1))$ we know that a tournament on n vertices with property S_k exists.

Dominating Sets

Let G = (V, E) be a graph. A set of vertices $D \subseteq V$ is called dominating with respect to G if every vertex in $V \setminus D$ is adjacent to a vertex in D.

Theorem: Suppose that $G = (V, E)$ is a graph with n vertices, and that $\delta(G) = \delta$, the minimum degree amongst G's vertices, is strictly positive. Then G contains a dominating set of size less than or equal to,

$$\frac{n \cdot (1 + \log(1 + \delta))}{1 + \delta}$$

Proof: Create a subset of G's vertices by choosing each $v \in V$ independently with probability p; call this subset X. Let Y be the collection of vertices in $V \setminus X$ without any neighbors in X; then, by definition, $X \cup Y$ is a dominating set for G.

What is the expected size of $|X \cup Y|$? Well; because they are disjoint subsets, we can calculate $|X \cup Y|$ by simply adding $|X|$ to $|Y|$:

$$\begin{aligned}
\mathbb{E}(|X|) &= \sum_{\upsilon\in V} \mathbb{E}\left(1_{\{\upsilon \text{ is chosen}\}}\right) \\
&= p \cdot n, \text{ while} \\
\mathbb{E}(|Y|) &= \sum_{\upsilon\in V} \mathbb{E}\left(1_{\{\upsilon \text{ is chosen}\}}\right) \\
&= \sum_{\upsilon\in V} \mathbb{E}\left(1_{\{\upsilon \text{ isn't in X, nor are any of its neighbors}\}}\right) \\
&= \sum_{\upsilon\in V} \mathbb{E}(1-p)^{\deg(\upsilon)+1}, \left(\text{b/c we've made } \deg(\upsilon)+1 \text{ choices independently}\right) \\
&\leq \sum_{\upsilon\in V} \mathbb{E}(1-p)^{\delta+1} \\
&= \sum_{\upsilon\in V} \mathbb{E}(1-p)^{\delta+1}
\end{aligned}$$

This tells us that,

$$\begin{aligned}
\mathbb{E}(|X \cup Y|) &\leq np + n(1-p)^{\delta+1} \\
&\leq np + ne^{-p(\delta+1)}
\end{aligned}$$

which has a minimum at,

$$p = \frac{\log(1+\delta)}{1+\delta}$$

Thus, for such p, we can find a dominating set of size at most,

$$\frac{n \cdot (1+\log(1+\delta))}{1+\delta},$$

as claimed.

The following question was first posed by Margulis: Given i.i.d random variables X, Y according to some distribution F, is there a constant C (independent of F; that is the important thing) such that,

$$\mathbb{P}(|X - Y| \leq 2) \leq C\mathbb{P}(|X - Y| \leq 1)?$$

It is far from obvious that such a $C < \infty$ must even exist. However, it is easy to see that such a C must be at least 3. Indeed, some X, Y are uniformly distributed on the even integers {2, 4, 2n} then it is easy to check that $\mathbb{P}(|X - Y| \leq 1) = 1/n$ and $\mathbb{P}(|X - Y| \leq 2) = \frac{3}{n} - \frac{2}{n^2}$. It was finally proved by Kozlov in the early 90s that the constant C = 3 actually works. Alon and Yuster shortly gave (at around the same time) another proof which was simpler and had the advantage that it actually established $\mathbb{P}(|X - Y| \leq r) < (2r - 1)\mathbb{P}(|X - Y| \leq 1)$, for any positive integer $r \geq 2$ which is also the best possible constant one can have for this inequality. We shall only show the weaker inequality with ≤ instead of the strict inequality. We shall later give mention briefly how one can improve the inequality to the strict inequality though we will not go over all the details.

Proof: The starting point for this investigation is based on one of the main tenets of Statistics: One can estimate (well enough) parametric information about a distribution from (large) finite samples from the same. In other words, if we wish to get more information about the unknown F, we could instead draw a large i.i.d sample $X_1, X_2, \ldots, X_m$ for a suitably large m and then the sample percentiles give information about F with high probability. This is in fact the basic premise of Non-parametric inference theory.

So, suppose we did draw such a large sample. Then a 'good' estimate for $\mathbb{P}(|X - Y| \leq 1)$ would be the ratio,

$$\frac{|\{(i,j) : |X_i - X_j| \leq 1\}|}{m^2}$$

A similar ratio, namely,

$$\frac{|\{(i,j) : |X_i - X_j| \leq 1\}|}{m^2}$$

should give a 'good' estimate for $\mathbb{P}(|X - Y| \leq r)$. This suggests the following question.

Suppose $T = (x_1, x_2, \ldots, x_m)$ is a sequence of (not necessarily distinct) reals, and $Tr := \{(i, j) : |x_i - x_j| \leq r\}$. Is it true that $|T_r| \leq (2r - 1)|T_1|$?

If this were false for some real sequence, one can consider F appropriately on the numbers in this sequence and maybe force a contradiction to the stated theorem. Thus, it behooves us to consider this (combinatorial) question posed above.

Let us try to prove the above by induction on m. For m = 1 there is nothing to prove. In fact, for m = 1 one in fact has strict inequality. So suppose we have (strict) inequality for r – 1 and we wish to prove the same for r.

Fix an i and let T o = T \ {x_i}. Consider the interval I := [x_i – 1, x_i + 1] and let S_I = {j|x_j ∈ I}, and let |S_I | = s. Then it is easy to see that,

$$T_1|=|T_1'|+(2s-1).$$

Now in order to estimate |T_r|, note that we need to estimate the number of pairs (j, i) such that |x_i – x_j | ≤ r. Suppose i was chosen such that |S_I | is maximum among all choices for x_i. Then observe that if we partition,

$$[x_i - r,\ x_i + r] = [x_i - r,\ x_i - (r-1))\cdots,\ [x_i - 2,\ x_1 - 1),$$
$$[x_i - 1,\ x_i + 1],(x_i + 1,\ x_i + 2],\cdots,(x_i + (r-1),\ x_i + r]$$

as indicated above, then in each of the intervals in this partition there are at most s values of j such that x_j is in that corresponding interval. This follows by the maximality assumption about x_i.

In fact, a moment's thought suggests a way in which this estimate can be improved. Indeed, if we also choose xi to be the largest among all xk that satisfy the previous criterion, then note that each of the intervals (xi + l, xi + (l + 1)] can in fact contain at most s – 1 xj 's. Thus it follows (by induction) that,

$$|T_r| \le T_r' + 1(r-1)s + (2s-1) + 2(r-1)(s-1) < (2r-1)T_1'| + (2r-1)(2s-1) = (2r-1)T_1|$$

This completes the induction and answers the question above, in the affirmative, with strict inequality.

Now, we are almost through. Suppose we do sample i.i.d observation $X_1, X_2, \ldots, X_m$ from the distribution F, and define the random variable $T_1 := |\{(i,j) : |X_i - X_j| \le 1\}|$ and $T_r := |\{(i,j) : |X_i - X_j| \le 1\}|$

$$\mathbb{E}(T_1) = \sum_{i \ne j} \mathbb{P}(|X_i - X_j| \le 1) + m = (m^2 - m)p_1 + m,$$

where $p_1 = \mathbb{P}(|X_i - X_j| \le 1)$. Similarly, we have,

$$\mathbb{E}(T - r) = (m^2 - m)p_r + m$$

with $p_r = \mathbb{P}(|X_i - X_j| \le r)$. By the inequality,

$$T_r < (2r-1)T_1$$

We have,

$$(m^2 - m)p_r + m = \mathbb{E}(T_r) < (2r-1)\mathbb{E}(T_1) = (2r-1)((m^2 - m)p_1 + m)$$

This simplifies to $p_r < (2r-1)p_1 + \frac{2r-2}{m-1}$. As $m \to \infty$, the desired result follows.

If $p_r = (2r-1)p_1$, then if we define $p_r(a) = \mathbb{P}(|X-a| \leq r)$ there exists some $a \in \mathbb{R}$ such that $p_r(a) > (2r-1)p_1(a)$. Once this is achieved, one can tweak the distribution F as follows.

Let X be a random variable that draws according to the distribution F with probability $1-\alpha$ and picks the number a (the one satisfying the inequality $p_r(a) > (2r-1)p_1(a)$) with probability α for a suitable α. Let us call this distribution G. Then from what we just proved above, it follows that $p^{(G)}r \leq (2r-1)p_1^{(G)}$. Here $p^{(G)}r$ denotes the probability $p_r = P(|X-Y| \leq r)$ if X, Y are picked i.i.d from the distribution G instead. However, if we calculate these terms, we see that $p^{(G)}r = p_r(1-\alpha)^2 + 2\alpha(1-\alpha)p_r(a) + \alpha^2$, so the above inequality reads,

$$p_r(1-\alpha)^2 + 2\alpha(1-\alpha)p_r(a) + \alpha^2 \leq (2_r - 1)\left(p_1(1-\alpha)^2 + 2\alpha(1-\alpha)p_1(a) + \alpha^2\right)$$

which holds if and only if $\alpha \geq \frac{\beta}{r-1+\beta}$ with $\beta = p_r(a) - (2r-1)p_1(a) > 0$. But since α is our choice, picking α smaller than this bound yields a contradiction.

References

- Combinatorics, science: britannica.com, Retrieved 28 March, 2020
- Basics-of-combinatorics, tutorial, combinatorics, math, practice: hackerearth.com, Retrieved 07 August, 2020
- Combinatorial-optimization, research: math.ethz.ch, Retrieved 16 April, 2020
- Combinatorial-geometry, combinatorics, science: britannica.com, Retrieved 09 June, 2020
- Probabilistic-Method, Probability: cut-the-knot.org, Retrieved 26 May, 2020
- The-Probabilistic-method-Combinatorics: math.iitb.ac.in, Retrieved 20 August, 2020

Theories in Combinatorics

CHAPTER 2

There are various theories that fall under combinatorics. Some of them are matroid theory, graph theory, order theory, discrete theory, combinatorial design theory, etc. This chapter closely examines these theories of combinatorics to provide an extensive understanding of the subject.

MATROID THEORY

A matroid is a structure that generalizes the properties of independence. Relevant applications are found in graph theory and linear algebra. There are several ways to define a matroid, each relate to the concept of independence.

Basic Linear Algebra

A is a 5×8 matrix, and its column vectors are in R^5. The set of column vectors of the matrix A are {1, 2, 3, 4, 5, 6, 7, 8}. We will focus on the set of column vectors in a matrix as the elements of a matroid.

$$
\begin{array}{c c}
 & \begin{array}{cccccccc} 1 & 2 & 3 & 4 & 5 & 6 & 7 & 8 \end{array} \\
A = & \begin{bmatrix} 0 & 1 & 1 & 1 & 0 & 0 & 0 & 0 \\ 0 & 1 & 1 & 0 & 1 & 1 & 0 & 0 \\ 0 & 0 & 0 & 1 & 1 & 0 & 1 & 0 \\ 0 & 0 & 0 & 0 & 0 & 1 & 1 & 1 \\ 0 & 0 & 0 & 0 & 0 & 0 & 0 & 1 \end{bmatrix}
\end{array}
$$

Now that we have a basic foundation of linear algebra and graph theory, we will begin our introduction of matroids by using the concept of a base.

Bases

A matroid M consists of a non-empty finite set E and a non-empty collection B of subsets of E, called bases, satisfying the following properties:

- B(i) no base properly contains another base.

- B(ii) if B_1 and B_2 are bases and if {e} is any element of B_1, then there is an element f of B_2 such that $(B_1 - \{e\}) \cup \{f\}$ is also a base.
- B(ii) is known as the exchange property: This property states that if an element is removed from B_1, then there exists an element in B_2, such that a new base, B_3, is formed when that element is added to B_1. We can use the property B(ii) to show that every base in a matroid has the same number of elements.

Theorem: Every base of a matroid has the same number of elements.

Proof: First assume that two bases of a matroid M, B_1 and B_2, contain different number of elements, such that $|B_1| < |B_2|$. Now suppose there is some element, $\{e_1\} \in M$, such that $e_1 \in B_1$, but $e_1 \notin B_2$. If we remove $\{e_1\}$ from B_1, then by B(ii), we know there is some element, $e_2 \in B_2$, but $e_2 \notin B_1$ such that $B_3 = B_1 \setminus (\{e_1\} \cup \{e_2\})$, where B_3 is a base in M. Therefore, $|B_1| = |B_3|$ but $|B_2| \neq |B_1| = |B_3|$.

If we continue the process of exchanging elements, defined by the property B9ii), k number of times, then there will be no element initially in B_1 that is not in the base B_k. Therefore, for all $e \in B_k$, the element e is also in B_2, and thus $B_k \subseteq B_2$.

From B(i), we know that no base properly contains another base. This is a contradiction. Therefore we know that every base has the same number of elements.

Example in Linear Algebra

Recall that in our previous example of the matrix A, the column vectors are in R^5. These columns form a matroid. We will take the base of a matroid to be a maximal linearly independent set that spans the column space (i.e., a basis for the column space). Consider two bases of our matroid:

$$B_1 = \left\{ \begin{bmatrix} 1 \\ 1 \\ 0 \\ 0 \\ 0 \end{bmatrix}, \begin{bmatrix} 1 \\ 0 \\ 1 \\ 0 \\ 0 \end{bmatrix}, \begin{bmatrix} 0 \\ 1 \\ 0 \\ 1 \\ 0 \end{bmatrix}, \begin{bmatrix} 0 \\ 0 \\ 0 \\ 1 \\ 1 \end{bmatrix} \right\}, \; B_2 = \left\{ \begin{bmatrix} 1 \\ 1 \\ 0 \\ 0 \\ 0 \end{bmatrix}, \begin{bmatrix} 0 \\ 1 \\ 1 \\ 0 \\ 0 \end{bmatrix}, \begin{bmatrix} 0 \\ 0 \\ 1 \\ 1 \\ 0 \end{bmatrix}, \begin{bmatrix} 0 \\ 0 \\ 0 \\ 1 \\ 1 \end{bmatrix} \right\},$$

Now if we remove the second vector in B1, then we can replace it with the second vector in B_2 to get a new base, B_3,

$$B_2 = \left\{ \begin{bmatrix} 1 \\ 1 \\ 0 \\ 0 \\ 0 \end{bmatrix}, \begin{bmatrix} 0 \\ 1 \\ 1 \\ 0 \\ 0 \end{bmatrix}, \begin{bmatrix} 0 \\ 1 \\ 0 \\ 1 \\ 0 \end{bmatrix}, \begin{bmatrix} 0 \\ 0 \\ 0 \\ 1 \\ 1 \end{bmatrix} \right\},$$

For this case, B(ii) is satisfied. We would find the same results if we continued this process with all possible bases of A. It is well known from Linear Algebra that no basis of A properly contains another basis.

Example in Graph Theory

We will take a base of our matroid to be a spanning tree of G. The following is a definition of a spanning tree.

Let G be a graph with n vertices. A spanning tree is a connected subgraph that uses all vertices of G that has n – 1 edges.

If we refer back to figure, then we can see that the bases of the graph, G, are in table.

Table: The Spanning Trees of G.

Bases
{a, b, c, d}
{a, e, d, c}
{b, c, d, e}
{b, a, e, d}
{c, b, a, e}
{c, b, f, e}
{c, d, f, a}
{c, g, a, e}
{c, g, f, e}

By observing the set of bases listed above, B(i) is satisfied, because no base properly contains another base. We can now demonstrate B(ii) by using this property with two bases. If we choose, B_1 = {a, b, c, d} and B_2 = {c, g, a, e}, then we can see the spanning trees of B_1 and B_2 in figures below.

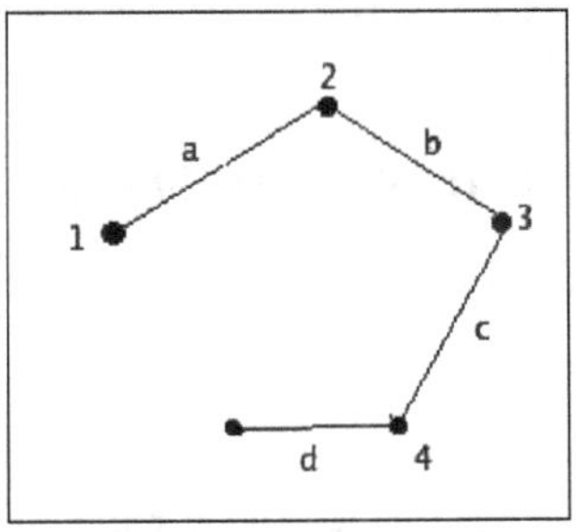

The Spanning Tree, B_1.

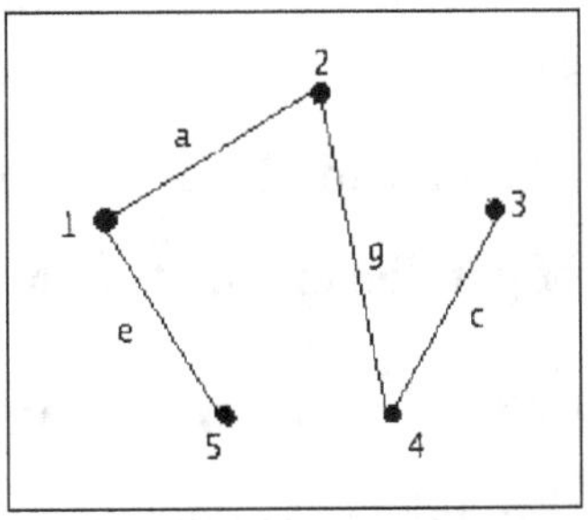

The Spanning Tree, B_2.

Each spanning tree has 5 vertices and 4 edges. We can demonstrate B(ii) by removing an element {a} from B_1, and then there exists an element in B_2 such that a new base is created, $B_3 = (B_1 \setminus \{a\}) \cup \{e\})$. Figure below shows the new base, B_3.

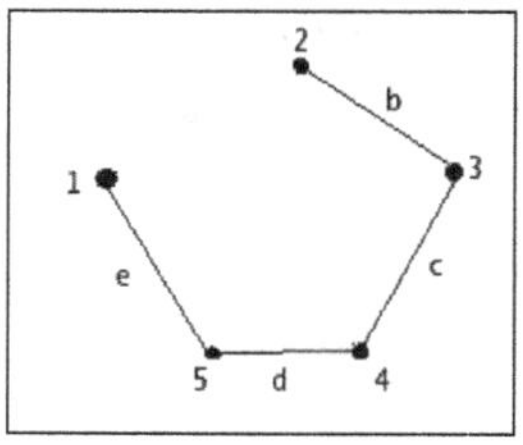

B_3.

A similar computation works for any choice of bases. Let T_1 and T_2 be spanning trees of a connected graph G:

- If e is any edge of T_1, show that there exists an edge f of T_2 such that the graph $(T_1 - \{e\}) \cup \{f\}$ (obtained from T_1 on replacing e by f) is also a spanning tree.
- Deduce that T_1 can be 'transformed' into T_2 by replacing the edges of T_1 one at a time by edges of T_2 in such a way that a spanning tree is obtained at each stage.

Because we take the spanning trees of a graph to be the bases of a matroid, we can conclude that the bases of a matroid have the same number of elements, and by the definition of a spanning tree has n – 1 elements (if there are n vertices).

Rank Function

A matroid consists of a non-empty finite set E and an integer-valued function r defined on the set of subset of E, satisfying:

- R(i) $0 \le r(A) \le |A|$, for each subset A of E;
- R(ii) if $A \subseteq B \subseteq E$, then $r(A) \le r(B)$;
- R(iii) for any $A, B \subseteq E$, $r(A \cup B) + r(A \cap B) \le r(A) + r(B)$.

The property R(i) guarantees that the rank of a subset cannot be negative, nor exceed its size. The second property guarantees that taking a superset does not decrease the rank of a set. The third property is equivalent to the exchange property that was defined in the previous section.

- A loop of a matroid M is an element e of E satisfying $r(\{e\}) = 0$.
- A pair of parallel elements of M is a pair {e, f} of E that satisfy $r\{e, f\} = 1$

Rank Function in Graph Theory

Recall that we can take the edges of a graph to be the elements of a matroid. For each subgraph, the rank will be the maximal number of edges in the subgraph that do not form a cycle.

We can show how the rank function works in graph theory using the following example. We will let E be the set of edges of the graph in figure. In figure, there are no cycles and the graph is connected. Therefore rank of A is the number of elements in A, so that r(A) = |A| = 2. Figure is the subgraph containing A = {c, d}.

In figure, there are four elements and one cycle. The rank of B is three, because the subsets of B with the maximum number of edges, which do not contain a cycle, are {b, c, d}, {b, c, e}, and {b, e, d}. Therefore, 3 = r(B) < |B| = 4.

The subset of E found in figure is a loop, with r(C) = 0. The subset of E in figure is a set of elements that are parallel elements. Therefore, r(D) = 1.

If we take the cycle, {c, d, e}, and remove any element of the cycle, the rank of the remaining elements will always be two, as shown in figure. Therefore, if we take the cycle with the remaining elements of E, we find that the rank of E is three, which means that the rank of the matroid is also three. The rank of M equals the size of a base of M.

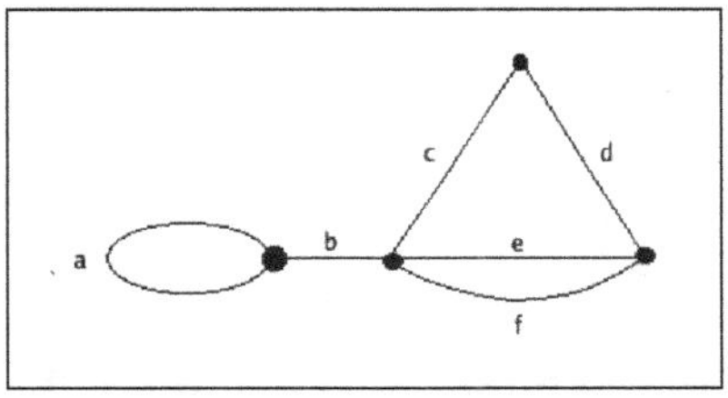

The set E.

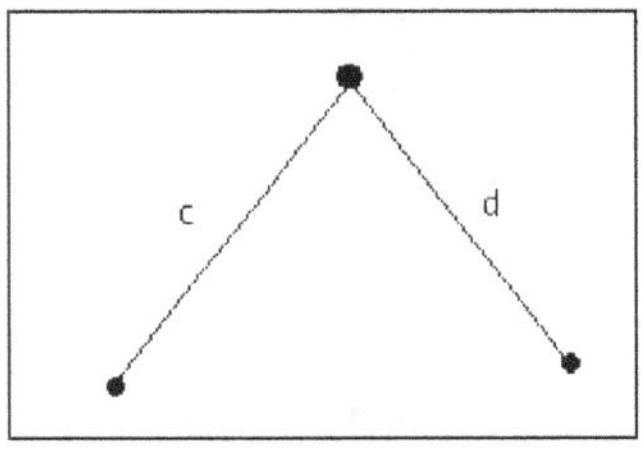

The subset {c, d} of E.

The rank of G, in figure, is 4, and because we take the set of edges of G as the elements of M, the rank of M is also r(M) = 4.

We can show an example of the property R(ii) in the graph G, by considering two subsets of G, A = {a, b, e} and B = {a, b, e, f}, so that A ⊆ B ⊆ E. In this case, r(A) = r(B) = 3. However, if we let C = {a, b, d, e}, then 3 = r(A) ≤ r(B) = 4. If we continue this example with other subsets, we would come to the same conclusion.

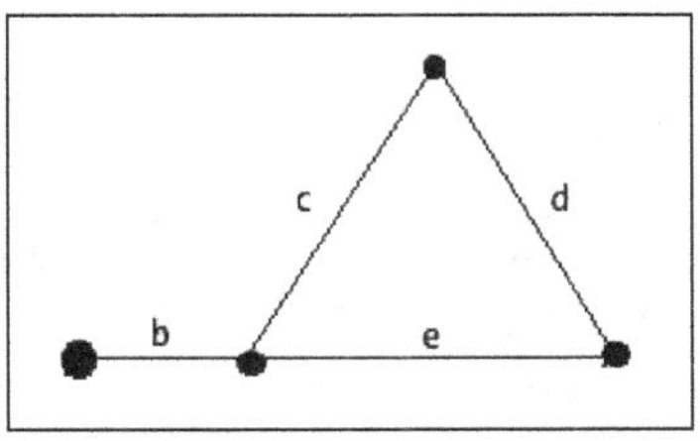

The subset {b, c, d, e} of E.

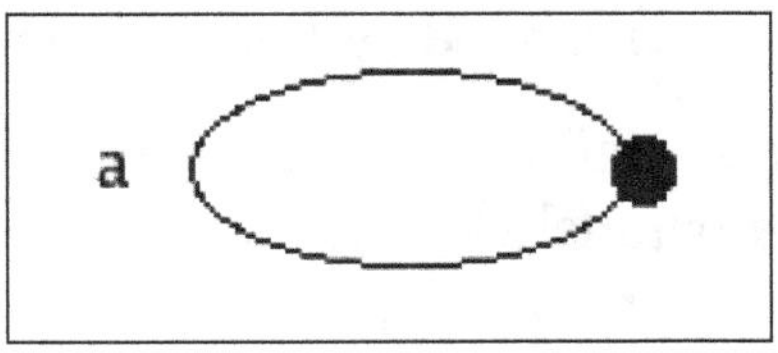

The subset C of E.

We can demonstrate the property R(iii) by using our previous example with two subsets of M being A = {a, b, e} and B = {a, b, d, e}.

$$r(A \cup c) + r(A \cap C) \leq r(A) + r(C)$$

$$r(\{a, b, d, e\}) + r(\{a, b.e\}) \leq r(\{a, b, d, e\}) + r(\{a, b, e\})$$

$$4+3<4+3$$

Therefore, property R(iii) is satisfied in this case.

Rank Function in Linear Algebra

We define rank(A) to be the size of a basis for span(A), or the dimension of the space spanned by A. Because we take the column vectors in a matrix to be the elements in our matroid, define our rank function to be the rank of each subset of M. In our previous example, one basis of the column space of A is,

$$B_1 = \left\{ \begin{bmatrix} 1 \\ 1 \\ 0 \\ 0 \\ 0 \end{bmatrix}, \begin{bmatrix} 1 \\ 0 \\ 1 \\ 0 \\ 0 \end{bmatrix}, \begin{bmatrix} 0 \\ 1 \\ 0 \\ 1 \\ 0 \end{bmatrix}, \begin{bmatrix} 0 \\ 0 \\ 0 \\ 1 \\ 0 \end{bmatrix} \right\},$$

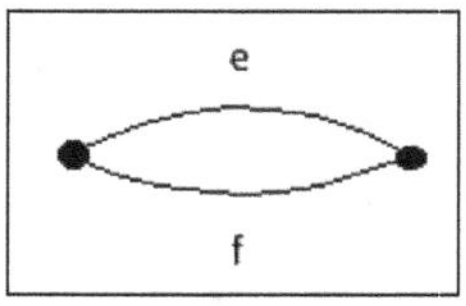

The subset D of E.

Because there are four column vectors in this basis, and this is a maximal linearly independent set, the rank of the matrix A is also four.

An example of a loop in a matrix is the zero column vector,

$$\begin{bmatrix} 0 \\ 0 \\ 0 \\ 0 \\ 0 \end{bmatrix}$$

because the space spanned by $\vec{0}$ is 0 dimensional.

The following is a definition of parallel elements in linear algebra. Two nonzero vectors, $\vec{u}$ and $\vec{v}$, are parallel elements, if $\vec{u} = \lambda\vec{v}$, for some scalar λ. An example of a set of parallel elements in a matrix is the set {e, f}, given by,

$$e = \begin{bmatrix} 1 \\ 0 \\ 1 \\ 0 \\ 0 \end{bmatrix} \quad f = \begin{bmatrix} 2 \\ 0 \\ 2 \\ 0 \\ 0 \end{bmatrix}$$

Because 2e = f, the rank of the set {e, f} is one. Therefore, We say that the set {e, f} is a set of parallel elements.

Now we can demonstrate the properties of a matroid in terms of its rank function by examining the matrix A. We can see the property, R(i), by observing the set, C, of column vectors from the matrix A.

$$C = \left\{ \begin{bmatrix} 1 \\ 1 \\ 0 \\ 0 \\ 0 \end{bmatrix}, \begin{bmatrix} 1 \\ 0 \\ 1 \\ 0 \\ 0 \end{bmatrix}, \begin{bmatrix} 0 \\ 1 \\ 0 \\ 1 \\ 0 \end{bmatrix}, \begin{bmatrix} 0 \\ 0 \\ 0 \\ 1 \\ 1 \end{bmatrix}, \begin{bmatrix} 0 \\ 1 \\ 1 \\ 0 \\ 0 \end{bmatrix} \right\},$$

In this example, the size of C is five, while the rank of C is four. Therefore, R(i) is satisfied for C is four. Therefore, R(i) is satisfied for C.

Now we will show an example of the property R(ii). If we continue with this example, and take,

$$D = \left\{ \begin{bmatrix} 1 \\ 1 \\ 0 \\ 0 \\ 0 \end{bmatrix}, \begin{bmatrix} 1 \\ 0 \\ 1 \\ 0 \\ 0 \end{bmatrix}, \begin{bmatrix} 0 \\ 1 \\ 0 \\ 1 \\ 0 \end{bmatrix} \right\},$$

so that $D \subseteq C \subseteq A$, we can see that $3 = r(D) < r(C) = 4$. Therefore, property R(ii) is satisfied in this example.

From the definition of a matroid, the equation in R(iii) is satisfied with the two subsets, $C, D \subset E$.

$$r(C \cup D) + r(C \cap D) \leq r(C) + r(D)$$

$$4 + 3 = 4 + 3$$

We would come to the same conclusion if we continued to examine various subsets of M.

Independent Sets

A subset of a matroid M is independent if it is contained in a base of a matroid. Conversely, a subset of M is dependent if it is not independent. We can also define a matroid in terms of its independent sets.

A matroid M consists of a non-empty finite set E and a non-empty collection I of subsets of E (called independent sets) satisfying the following properties:

- I(i) any subset of an independent set is independent; I(ii) if I and J are independent sets with |J| > |I|, then there is an element e, contained in J but not in I, such that I ∪ {e} is independent.

- To explain property I(i), we will let K be a subset of a non-empty finite set E. We know that if K is independent, then it is contained in a base. Therefore, any subset of K is independent because the subset is also contained within a base.
- Property I(ii) is the equivalent of the exchange axiom, which was defined in the section on bases. This property states that if two independents sets satisfy the inequality |J| > |I|, then there exists an element in J, such that the new independent set is formed when that element is added to I.

The connection to the previous sections. Moreover, if A is an independent set, then A is contained in some base of M, which implies that r(A) = |A|.

Independent Sets in Graph Theory

We will take the independent sets of a graph to be the sets of edges in a graph that do not contain a cycle. Recall that in graph theory, a cycle is a closed path. Another definition of independent sets in graph theory uses forests.

A forest is a graph that contains no cycles. A connected forest is a tree. We can say that the independent sets of a graph are the edge sets of the forests contained in the graph. Figures below are examples of forests contained in the graph G.

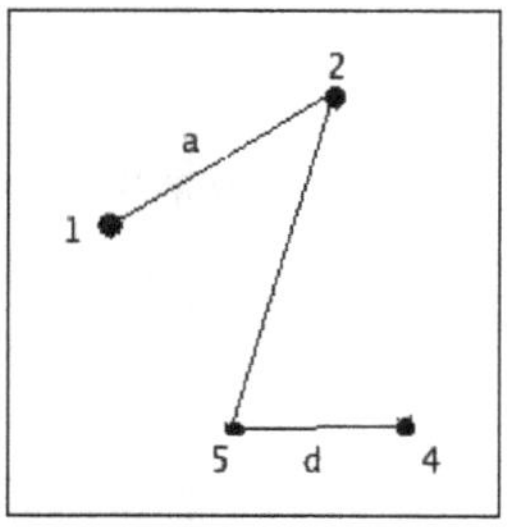

An Example of a Forest Contained in G.

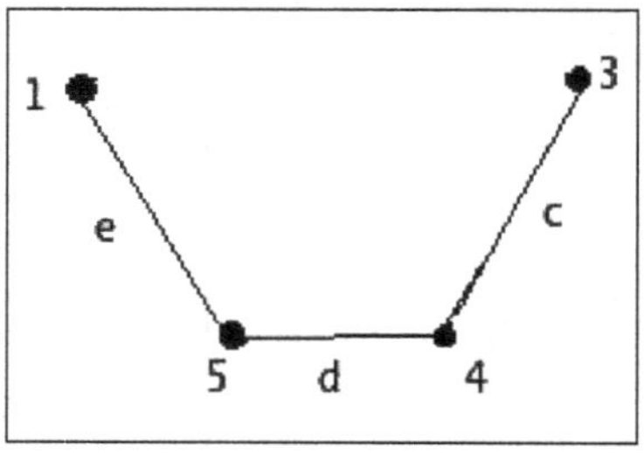

Another Example of a Forest Contained in G.

The first property I(i), can be shown because a set is independent if it is contained within a base. Therefore independent sets must be contained within a spanning tree of a graph, which means that the rank of an independent set must be less than or equal to the rank of the graph. Table lists the forests contained in the graph G defined in figure.

Table: The Forests of G.

{a}, {b}, {c}, {d}, {e}
{f}, {g}, {a, b}, {b, c}
{c, d}, {d, e}, {e, f}, {f, g}
{g, a}, {a, f}, {e, f}, {d, f}
{b, f}, {b, g}, {c, g}, {d, g}

{a, b, c}, {a, b, g}, {a, e, d}
{a, f, d}, {a, g, c}, {a, g, d}
{b, c, d}, {b, g, d}, {b, f, d}
{b, f, e}, {c, d, e}, {c, g, d}
{c, d, f}, {e, f, e}, {a, f, g}
{a, b, c, d}, {a, e, d, c}, {b, c, d, e}
{b, a, e, d}, {c, b, a, e}, {c, b, f, e}
{c, d, f, a}, {c, g, a, e}, {c, g, f, e}

From observing the table of forests, we can see that the forests are contained within the spanning trees, which are the bases listed in the last three rows.

Now we will demonstrate why the exchange axiom for independent sets requires that two independent sets, K and L, must satisfy the inequality $|K| > |L|$. Suppose we let the two forests contained in G be the sets K and L_0 shown in figures. Notice that $|K| = |L| = 3$.

We find that there is no element, e contained in K but not L, such that the set $L \cup \{e\}$ is independent.

However, if we let $L_1 = L_0 \setminus \{c\}$, so that $3 = |K| > |L_1| = 2$, then we necessarily have an element, in this case {d}, such that $d \in K$ but not in L1. Therefore, we find the independent set $L_1 \cup \{d\}$.

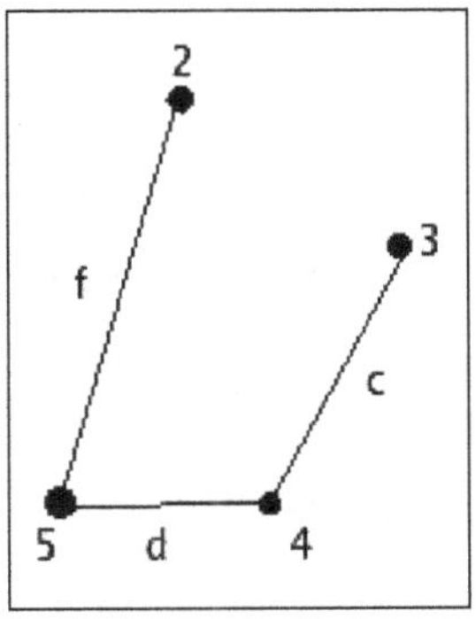

The Forest, K.

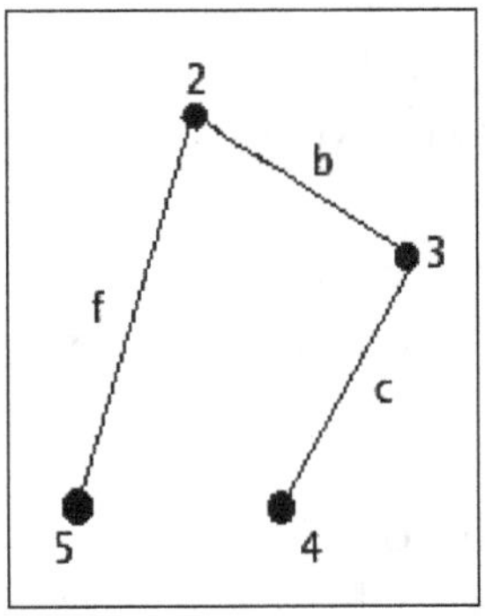

The Forest, L_0.

Independent Sets in Linear Algebra

We will take the independent sets of a matroid, M, of column vectors. I is independent in M if I is linearly independent.

Linear Algebra and its Applications

An indexed set of vectors $\{v_1,.., v_p\}$ in R_n is said to be linearly independent if the vector equation,

$$x_1 v_1 + x_2 v_2 + \dots + x_p v_p = 0$$

has only the trivial solution. The set $\{v_1 ..., v_p\}$ is said to be linearly dependent if there exists weights $c_1,..., c_p$, not all zero, such that,

$$c_1 v_1 + c_2 v_2 + ...c_p v_p = 0$$

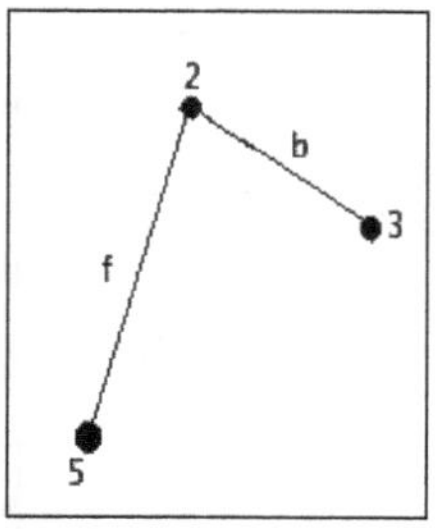

The Forest, L_1.

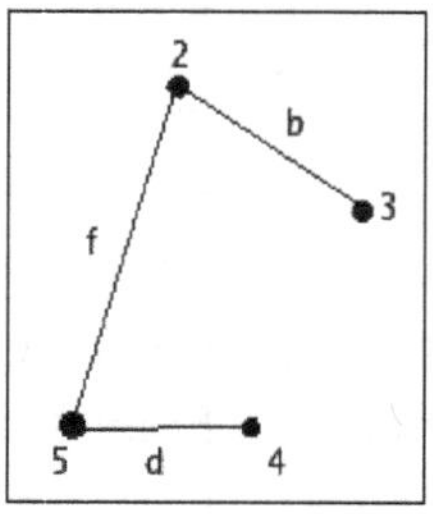

The Forest, $L_1 \cup \{d\}$.

Therefore, if we take B_1 as a base of the matrix A, we know that a base is defined as a maximal linearly independent set that spans the column space. Therefore, if we take B_1 as a base of the matrix A, namely then any combination of the column vectors would create an independent set.

$$B_1 = \left\{ \begin{bmatrix} 1 \\ 1 \\ 0 \\ 0 \\ 0 \end{bmatrix}, \begin{bmatrix} 1 \\ 0 \\ 1 \\ 0 \\ 0 \end{bmatrix}, \begin{bmatrix} 0 \\ 1 \\ 0 \\ 1 \\ 0 \end{bmatrix}, \begin{bmatrix} 0 \\ 0 \\ 0 \\ 1 \\ 1 \end{bmatrix} \right\},$$

To show that B_1 is a linearly independent set of vectors, and we can take the set of vectors, and designate them as the column vectors in the matrix, b_1,

$$\begin{bmatrix} 1 & 0 & 0 & 0 \\ 0 & 1 & 0 & 0 \\ 0 & 0 & 1 & 0 \\ 0 & 0 & 0 & 1 \\ 0 & 0 & 0 & 0 \end{bmatrix}$$

Now we will take the column vectors as the set of vectors in B_2.

$$B_2 = \left\{ \begin{bmatrix} 1 \\ 1 \\ 0 \\ 0 \\ 0 \end{bmatrix}, \begin{bmatrix} 0 \\ 1 \\ 1 \\ 0 \\ 0 \end{bmatrix}, \begin{bmatrix} 0 \\ 0 \\ 1 \\ 1 \\ 0 \end{bmatrix}, \begin{bmatrix} 0 \\ 0 \\ 0 \\ 1 \\ 0 \end{bmatrix} \right\},$$

The following examples are subsets of B_2.

$$c_1 \begin{bmatrix} 0 \\ 0 \\ 0 \\ 0 \\ 0 \end{bmatrix} + c_2 \begin{bmatrix} 0 \\ 0 \\ 0 \\ 1 \\ 0 \end{bmatrix} + c_3 \begin{bmatrix} 0 \\ 0 \\ 0 \\ 0 \\ 1 \end{bmatrix} = 0$$

$$b_1 \begin{bmatrix} 0 \\ 1 \\ 0 \\ 0 \\ 0 \end{bmatrix} + b_2 \begin{bmatrix} 0 \\ 0 \\ 1 \\ 0 \\ 0 \end{bmatrix} = 0$$

In these two examples, we find that $b_1 = b_2 = 0$ and $c_1 = c_2 = c_3 = 0$ are the only solutions. Therefore, both subsets are linearly independent.

To demonstrate I(i) and I(ii), we can take two independent sets of the matrix A to be K and L,

$$K = \left\{ \begin{bmatrix} 1 \\ 0 \\ 1 \\ 0 \\ 0 \end{bmatrix}, \begin{bmatrix} 0 \\ 1 \\ 0 \\ 1 \\ 0 \end{bmatrix}, \begin{bmatrix} 0 \\ 0 \\ 0 \\ 1 \\ 1 \end{bmatrix} \right\},$$

$$L = \left\{ \begin{bmatrix} 1 \\ 1 \\ 0 \\ 0 \\ 0 \end{bmatrix}, \begin{bmatrix} 0 \\ 0 \\ 0 \\ 1 \\ 1 \end{bmatrix} \right\},$$

so that $|K| > |L|$. From our previous discussion of linear independence in a set of vectors, we can see that any subset of K or L are linearly independent.

The inequality stated in I(ii) ensures that the dimension of the space spanned by K is greater than the dimension of the space spanned by L, which makes it impossible to add an element from K to L. For example, given three vectors that span a space, we can extend a different set of two vectors which span a plane to a set of three vectors which spans a space.

Now we can demonstrate the exchange property by noticing that K and L share a common element,

$$\begin{bmatrix} 0 \\ 0 \\ 0 \\ 1 \\ 1 \end{bmatrix};$$

which means that we must choose one vector from the set to add to L, so that L is independent.

$$\left\{ \begin{bmatrix} 1 \\ 0 \\ 1 \\ 0 \\ 0 \end{bmatrix}, \begin{bmatrix} 0 \\ 1 \\ 0 \\ 1 \\ 0 \end{bmatrix} \right\}$$

We can see the linear independence in the sets;

$$L_1 = \left\{ \begin{bmatrix} 1 \\ 1 \\ 0 \\ 0 \\ 0 \end{bmatrix}, \begin{bmatrix} 1 \\ 0 \\ 1 \\ 0 \\ 0 \end{bmatrix}, \begin{bmatrix} 0 \\ 0 \\ 0 \\ 1 \\ 1 \end{bmatrix} \right\} \text{ and}$$

$$L_2 = \left\{ \begin{bmatrix} 0 \\ 1 \\ 0 \\ 1 \\ 0 \end{bmatrix}, \begin{bmatrix} 1 \\ 0 \\ 1 \\ 0 \\ 0 \end{bmatrix}, \begin{bmatrix} 0 \\ 0 \\ 0 \\ 1 \\ 1 \end{bmatrix} \right\}$$

A matroid M consists of a non-empty finite set E, and a collection C of non-empty subsets of E (called cycles) satisfying the following properties:

- C(i) no cycle properly contains another cycle; C(ii) if C_1 and C_2 are two distinct cycles each containing an element e, then there exists a cycle in $C_1 \cup C_2$ that does not contain {e}.

Now we can connect a cycle to the concepts introduced in the previous sections. Let A be a cycle. A – {e} is in some base for all e ∈ A, which implies that r(A) = |A| – 1, and r(B) = |B| for all B ⊂ A. Therefore, A is minimally dependent, which means that if we take any element from A, then the remaining set is linearly independent.

Cycles in Graph Theory

A cycle is a minimally dependent set, which means that any element can be removed from the set, and the set will become independent. This property can be seen in figures. The graph in figure shows a set that is dependent, but which is not minimally dependent. There exists an element, {b}, which can be removed while a cycle still exists in the set.

We can also define a cycle in graph theory in terms of a path, and so we will take a cycle of a matroid, M, to be a closed path of G containing at least one edge.

The cycles of the graph G in figure are provided in the table:

Cycles
{a, b, c, d, e}
{a, e, f}
{a, e, d, g}
{d, f, g}
{b, c, g}
{b, c, d, f}

Two graphic examples of cycles found in figure:

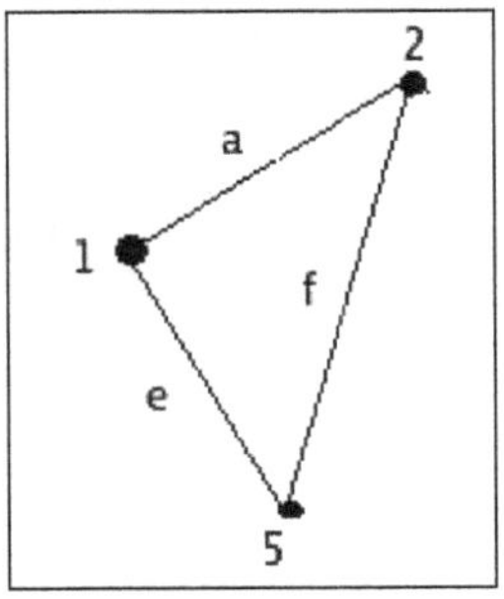

The Cycle, C_1.

The property C(i) holds by observing the table of cycles in G.

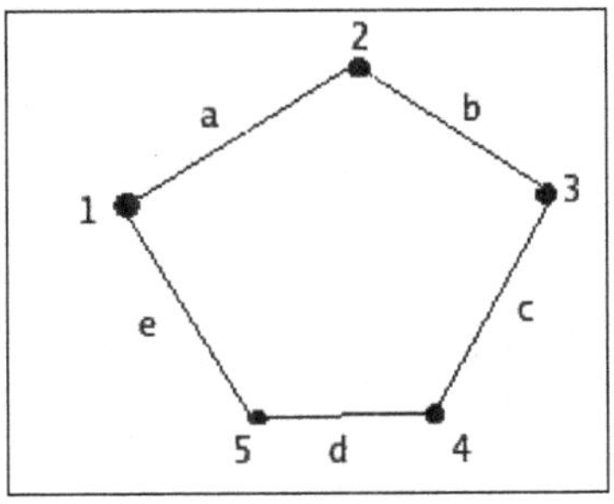

The Cycle, C_2.

Using our examples of the cycles C_1 and C_2, we can see that the two cycles each contain the elements {a} and {e}. Figure shows the graph of $C_3 = C_1 \cup C_2$. We can see there are three cycles in C_3, which are {a, e, f}, {a, b, c, d, e}, and {b, c, d, f}.

The cycle {b, c, d, f} in figure is a cycle in C_3 which contains neither {a} nor {e}. Therefore, the property C(ii) holds in this case.

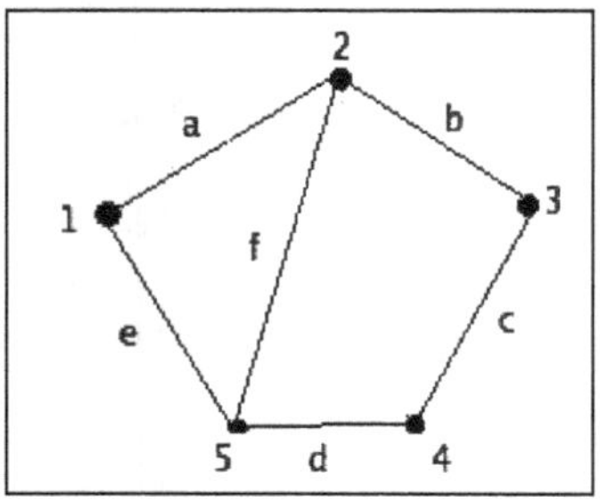

The Cycles, C_3.

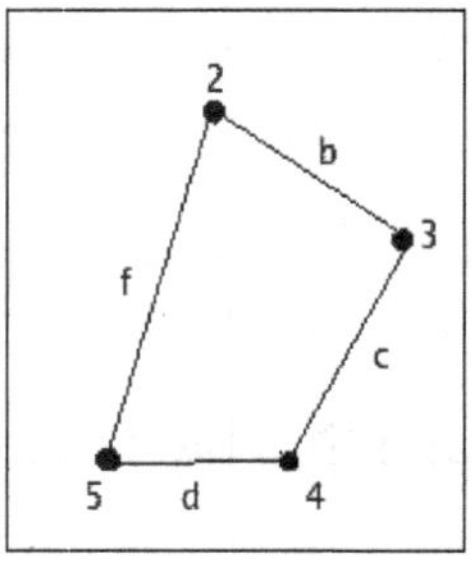

The Cycle, C_4.

Cycles in Linear Algebra

We will take the cycles of a matrix to be a set of minimally dependent column vectors. To show examples of cycles in linear algebra, we can take L_1 and L_2 to be two cycles in the matrix, A.

$$L_1 = \left\{ \begin{bmatrix} 1 \\ 1 \\ 0 \\ 0 \\ 0 \end{bmatrix}, \begin{bmatrix} 1 \\ 0 \\ 1 \\ 0 \\ 0 \end{bmatrix}, \begin{bmatrix} 1 \\ 1 \\ 1 \\ 0 \\ 0 \end{bmatrix}, \begin{bmatrix} 0 \\ 1 \\ 1 \\ 0 \\ 0 \end{bmatrix} \right\} \text{ and}$$

$$L_2 = \left\{ \begin{bmatrix} 0 \\ 1 \\ 1 \\ 0 \\ 0 \end{bmatrix}, \begin{bmatrix} 0 \\ 1 \\ 0 \\ 1 \\ 0 \end{bmatrix}, \begin{bmatrix} 0 \\ 1 \\ 1 \\ 1 \\ 0 \end{bmatrix}, \begin{bmatrix} 0 \\ 0 \\ 1 \\ 1 \\ 0 \end{bmatrix} \right\}.$$

We will see L_1 and L_2 are a minimally dependent set of column vectors.

$$d_1 \begin{bmatrix} 1 \\ 1 \\ 0 \\ 0 \\ 0 \end{bmatrix} + d_2 \begin{bmatrix} 1 \\ 0 \\ 1 \\ 0 \\ 0 \end{bmatrix} + d_3 \begin{bmatrix} 1 \\ 1 \\ 1 \\ 0 \\ 0 \end{bmatrix} + d_4 \begin{bmatrix} 0 \\ 1 \\ 1 \\ 0 \\ 0 \end{bmatrix} = 0$$

For L_1, we can let $d_1 = d_2 = d_4 = 1$ and $d_3 = -2$, so that the column vectors of L_1 add to the zero vector. For L_2, we can let $e_1 = e_2 = e_4 = 1$ and $e_3 = -2$, so that the column vectors of L_2 add to the zero vector. L_1 and L_2 are linearly dependent. Notice that if you remove any column vector from L_1 or L_2, then the set is linearly independent. Therefore, L_1 and L_2 are cycles.

To demonstrate C(ii), we will let $L_3 = L_1 \cup L_2$.

$$L_3 \left\{ \begin{bmatrix} 1 \\ 1 \\ 0 \\ 0 \\ 0 \end{bmatrix}, \begin{bmatrix} 1 \\ 0 \\ 1 \\ 0 \\ 0 \end{bmatrix}, \begin{bmatrix} 0 \\ 1 \\ 1 \\ 0 \\ 0 \end{bmatrix}, \begin{bmatrix} 0 \\ 1 \\ 0 \\ 1 \\ 0 \end{bmatrix}, \begin{bmatrix} 0 \\ 0 \\ 1 \\ 1 \\ 0 \end{bmatrix} \right\},$$

The common column vector of L_1 and L_2 is,

$$f = \begin{bmatrix} 0 \\ 1 \\ 1 \\ 0 \\ 0 \end{bmatrix}$$

Therefore, there is a cycle contained in L_3 that does not contain f, namely the set of column vectors.

$$L_4 \left\{ \begin{bmatrix} 1 \\ 1 \\ 0 \\ 0 \\ 0 \end{bmatrix}, \begin{bmatrix} 1 \\ 0 \\ 1 \\ 0 \\ 0 \end{bmatrix}, \begin{bmatrix} 0 \\ 1 \\ 0 \\ 1 \\ 0 \end{bmatrix}, \begin{bmatrix} 0 \\ 0 \\ 1 \\ 1 \\ 0 \end{bmatrix} \right\},$$

Therefore, the property C(ii) is satisfied for this example.

Vertex-edge Incidence Matrix

Thus far, we have seen how both graphs and matrices can be viewed as matroids. Now we will link graph theory and linear algebra by translating a graph to a unique matrix, and vice versa, using the language of matroids. The vertex-edge incidence matrix demonstrates the relationship between a matrix and a graph.

Theorem: Let G be a graph and AG be its vertex-edge incidence matrix. When the entries of A_G are viewed modulo(2), its vector matroid $M[A_G]$ has as its independent sets all subsets of E(G) that do not contain the edges of a cycle. Thus $M[A_G] = M(G)$ and every graphic matroid is binary.

The idea of a matrix being viewed mod(2), means that the entries of the matrix are either 0 or 1. Since the vertex-edge incidence matrix represents a graph, we call the graph binary. Because we have shown that a graph can be viewed as a matroid, we can say that the matroid is also binary.

The following is an example of a vertex-edge incidence matrix using the graph in figure. If an edge and a vertex are incident on a graph, then the corresponding entry in the matrix is 1. Otherwise, if an edge and vertex are not incident, then the corresponding entry in the matrix is zero.

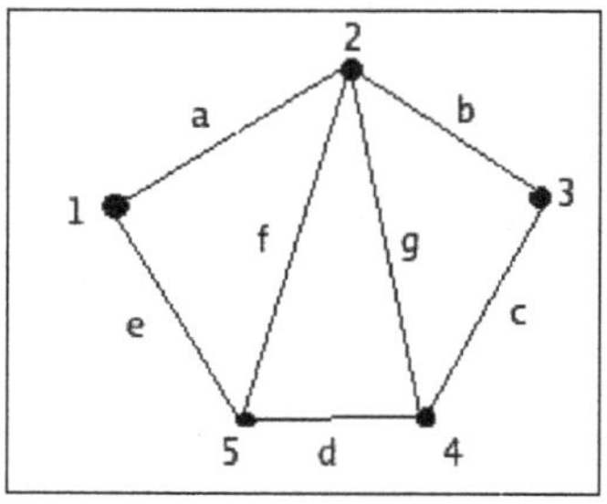

Graph, G.

$$\begin{array}{c} \\ 1 \\ 2 \\ 3 \\ 4 \\ 5 \end{array}\begin{array}{c} \begin{array}{ccccccc} a & b & c & d & e & f & g \end{array} \\ \begin{bmatrix} 1 & 0 & 0 & 0 & 1 & 0 & 0 \\ 1 & 1 & 0 & 0 & 0 & 1 & 1 \\ 0 & 1 & 1 & 0 & 0 & 0 & 0 \\ 0 & 0 & 1 & 1 & 0 & 0 & 1 \\ 0 & 0 & 0 & 1 & 1 & 1 & 0 \end{bmatrix} \end{array}$$

We can see the relationship between graphs and matrices in the vertex-edge incidence matrix if we use the set {a, e, f} as our example. We can see that the rank of the set {a, e, f} is 2, because any subset, containing two elements, does not contain a cycle. In the graph in figure, we can see that the set {a, e, f} is a cycle. The sum of the column vectors corresponding to the set of edges in our example is,

$$\begin{bmatrix} 1 \\ 1 \\ 0 \\ 0 \\ 0 \end{bmatrix} + \begin{bmatrix} 1 \\ 0 \\ 0 \\ 0 \\ 1 \end{bmatrix} + \begin{bmatrix} 0 \\ 1 \\ 0 \\ 0 \\ 1 \end{bmatrix} = \begin{bmatrix} 0 \\ 0 \\ 0 \\ 0 \\ 0 \end{bmatrix}.$$

This set of vectors is minimally dependent. If you take any one vector from the set, the set become an independent set.

The rank of the column vectors corresponding to the set {a, e, f} is also two, because any subset of the set of column vectors, containing two elements, does not contain a cycle.

One base of G is the set of edges {a, b, c, d}. The corresponding set of column vectors are,

$$N = \left\{ \begin{bmatrix} 1 \\ 1 \\ 0 \\ 0 \\ 0 \end{bmatrix}, \begin{bmatrix} 0 \\ 1 \\ 1 \\ 0 \\ 0 \end{bmatrix}, \begin{bmatrix} 0 \\ 0 \\ 1 \\ 1 \\ 0 \end{bmatrix}, \begin{bmatrix} 0 \\ 0 \\ 0 \\ 1 \\ 1 \end{bmatrix} \right\}.$$

The set of vectors are a maximal independent set, because $|N| = r(N) = 4$. Therefore, the set of vectors, N, is also a base.

Now we can see the link between graph theory and linear algebra, by using the language of matroids to motivate our discussion and to generalize the properties of independence.

GRAPH THEORY

In mathematics, graph theory is the study of graphs, which are mathematical structures used to model pairwise relations between objects. A graph in this context is made up of vertices (also called nodes or points) which are connected by edges (also called links or lines). A distinction is made between undirected graphs, where edges link two vertices symmetrically, and directed graphs, where edges link two vertices asymmetrically;

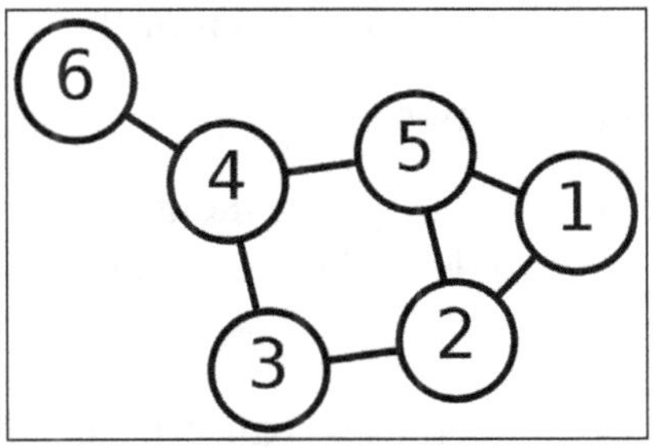

A drawing of a graph.

Definitions in graph theory vary. The following are some of the more basic ways of defining graphs and related mathematical structures.

Graph

In one restricted but very common sense of the term, a graph is an ordered pair G = (V, E) comprising:

- V a set of vertices (also called nodes or points).

- $E \subseteq \{\{x, y\} \mid (x, y) \in V^2 \wedge x \neq y\}$ a set of edges (also called links or lines), which are unordered pairs of vertices (i.e., an edge is associated with two distinct vertices).

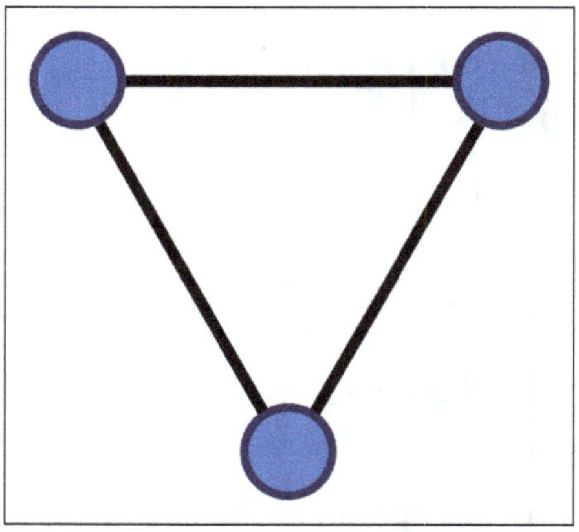

A graph with three vertices and three edges.

To avoid ambiguity, this type of object may be called precisely an undirected simple graph.

In the edge $\{x, y\}$, the vertices x and y are called the endpoints of the edge. The edge is said to join x and y and to be incident on x and on y. A vertex may exist in a graph and not belong to an edge. Multiple edges are two or more edges that join the same two vertices.

In one more general sense of the term allowing multiple edges, a graph is an ordered triple $G = (V, E, \phi)$ comprising:

- V a set of vertices (also called nodes or points);
- E a set of edges (also called links or lines);
- $\phi: E \rightarrow \{\{x, y\} \mid (x, y) \in V^2 \wedge x \neq y\}$ an incidence function mapping every edge to an unordered pair of vertices (i.e., an edge is associated with two distinct vertices).

To avoid ambiguity, this type of object may be called precisely an undirected multigraph.

A loop is an edge that joins a vertex to itself. Graphs as defined in the two definitions above cannot have loops, because a loop joining a vertex x is the edge (for an undirected simple graph) or is incident on (for an undirected multigraph) $\{x, x\} = \{x\}$ which is not in $\{\{x, y\} \mid (x, y) \in V^2 \wedge x \neq y\}$. So to allow loops the definitions must be expanded. For undirected simple graphs, $E \subseteq \{\{x, y\} \mid (x, y) \in V^2 \wedge x \neq y\}$ should become $E \subseteq \{\{x, y\} \mid (x, y) \in V^2\}$. For undirected multigraphs, $\phi: E \rightarrow \{\{x, y\} \mid (x, y) \in V^2 \wedge x \neq y\}$ should become $\phi: E \rightarrow \{\{x, y\} \mid (x, y) \in V^2\}$. To avoid ambiguity, these types of objects may be called precisely an undirected simple graph permitting loops and an undirected multigraph permitting loops respectively.

V and E are usually taken to be finite, and many of the well-known results are not true (or are rather different) for infinite graphs because many of the arguments fail in the infinite case. Moreover, V is often assumed to be non-empty, but E is allowed to be the empty set. The order of a graph is $|V|$, its number of vertices. The size of a graph is $|E|$, its number of edges. The degree or valency of a vertex is the number of edges that are incident to it, where a loop is counted twice.

In an undirected simple graph of order n, the maximum degree of each vertex is $n - 1$ and the maximum size of the graph is $n(n - 1)/2$.

The edges of an undirected simple graph permitting loops G induce a symmetric homogeneous relation ~ on the vertices of G that is called the adjacency relation of G. Specifically, for each edge {x, y}, its endpoints x and y are said to be adjacent to one another, which is denoted $x \sim y$.

Directed Graph

A directed graph or digraph is a graph in which edges have orientations.

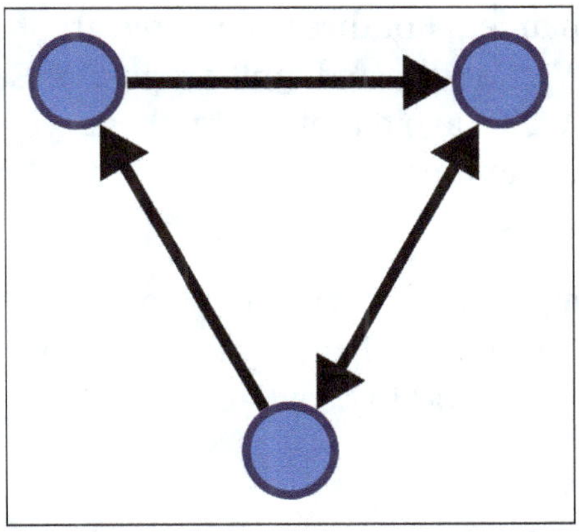

A directed graph with three vertices and four directed edges (the double arrow represents an edge in each direction).

In one restricted but very common sense of the term, a directed graph is an ordered pair G = (V, E) comprising:

- V a set of vertices (also called nodes or points);
- $E \subseteq \{(x, y) \mid (x, y) \in V^2 \wedge x \neq y\}$ a set of edges (also called directed edges, directed links, directed lines, arrows or arcs) which are ordered pairs of distinct vertices (i.e., an edge is associated with two distinct vertices).

To avoid ambiguity, this type of object may be called precisely a directed simple graph.

In the edge (x, y) directed from x to y, the vertices x and y are called the endpoints of the edge, x the tail of the edge and y the head of the edge. The edge (y, x) is called the inverted edge of (x, y). The edge is said to join x and y and to be incident on x and on

y. A vertex may exist in a graph and not belong to an edge. A loop is an edge that joins a vertex to itself. Multiple edges are two or more edges that join the same two vertices.

In one more general sense of the term allowing multiple edges, a directed graph is an ordered triple G = (V, E, ϕ) comprising:

- V a set of vertices (also called nodes or points);
- E a set of edges (also called directed edges, directed links, directed lines, arrows or arcs);
- $\phi: E \rightarrow \{(x, y) \mid (x, y) \in V^2 \wedge x \neq y\}$ an incidence function mapping every edge to an ordered pair of distinct vertices (i.e., an edge is associated with two distinct vertices).

To avoid ambiguity, this type of object may be called precisely a directed multigraph.

Directed graphs as defined in the two definitions above cannot have loops, because a loop joining a vertex x is the edge (for a directed simple graph) or is incident on (for a directed multigraph) (x, x) which is not in $\{(x, y) \mid (x, y) \in V^2 \wedge x \neq y\}$. So to allow loops the definitions must be expanded. For directed simple graphs, $E \subseteq \{(x, y) \mid (x, y) \in V^2 \wedge x \neq y\}$ should become $E \subseteq V^2$. For directed multigraphs, $\phi: E \rightarrow \{(x, y) \mid (x, y) \in V^2 \wedge x \neq y\}$ should become $\phi: E \rightarrow V^2$. To avoid ambiguity, these types of objects may be called precisely a directed simple graph permitting loops and a directed multigraph permitting loops (or a quiver) respectively.

The edges of a directed simple graph permitting loops G is a homogeneous relation ~ on the vertices of G that is called the adjacency relation of G. Specifically, for each edge (x, y), its endpoints x and y are said to be adjacent to one another, which is denoted x ~ y.

Graph Drawing

Graphs are represented visually by drawing a point or circle for every vertex, and drawing a line between two vertices if they are connected by an edge. If the graph is directed, the direction is indicated by drawing an arrow.

A graph drawing should not be confused with the graph itself (the abstract, non-visual structure) as there are several ways to structure the graph drawing. All that matters is which vertices are connected to which others by how many edges and not the exact layout. In practice, it is often difficult to decide if two drawings represent the same graph. Depending on the problem domain some layouts may be better suited and easier to understand than others.

The pioneering work of W. T. Tutte was very influential on the subject of graph drawing. Among other achievements, he introduced the use of linear algebraic methods to obtain graph drawings.

Graph drawing also can be said to encompass problems that deal with the crossing number and its various generalizations. The crossing number of a graph is the minimum number of intersections between edges that a drawing of the graph in the plane must contain. For a planar graph, the crossing number is zero by definition.

Graph-theoretic Data Structures

There are different ways to store graphs in a computer system. The data structure used depends on both the graph structure and the algorithm used for manipulating the graph. Theoretically one can distinguish between list and matrix structures but in concrete applications the best structure is often a combination of both. List structures are often preferred for sparse graphs as they have smaller memory requirements. Matrix structures on the other hand provide faster access for some applications but can consume huge amounts of memory. Implementations of sparse matrix structures that are efficient on modern parallel computer architectures are an object of current investigation.

List structures include the incidence list, an array of pairs of vertices, and the adjacency list, which separately lists the neighbors of each vertex: Much like the incidence list, each vertex has a list of which vertices it is adjacent to.

Matrix structures include the incidence matrix, a matrix of 0's and 1's whose rows represent vertices and whose columns represent edges, and the adjacency matrix, in which both the rows and columns are indexed by vertices. In both cases a 1 indicates two adjacent objects and a 0 indicates two non-adjacent objects. The degree matrix indicates the degree of vertices. The Laplacian matrix is a modified form of the adjacency matrix that incorporates information about the degrees of the vertices, and is useful in some calculations such as Kirchhoff's theorem on the number of spanning trees of a graph. The distance matrix, like the adjacency matrix, has both its rows and columns indexed by vertices, but rather than containing a 0 or a 1 in each cell it contains the length of a shortest path between two vertices.

Problems

Enumeration

There is a large literature on graphical enumeration: the problem of counting graphs meeting specified conditions.

Subgraphs, Induced Subgraphs and Minors

A common problem, called the subgraph isomorphism problem, is finding a fixed graph as a subgraph in a given graph. One reason to be interested in such a question is that many graph properties are hereditary for subgraphs, which means that a graph has the property if and only if all subgraphs have it too. Unfortunately, finding maximal subgraphs of a certain kind is often an NP-complete problem. For example:

- Finding the largest complete subgraph is called the clique problem (NP-complete).

One special case of subgraph isomorphism is the graph isomorphism problem. It asks whether two graphs are isomorphic. It is not known whether this problem is NP-complete, nor whether it can be solved in polynomial time.

A similar problem is finding induced subgraphs in a given graph. Again, some important graph properties are hereditary with respect to induced subgraphs, which means that a graph has a property if and only if all induced subgraphs also have it. Finding maximal induced subgraphs of a certain kind is also often NP-complete. For example:

- Finding the largest edgeless induced subgraph or independent set is called the independent set problem (NP-complete).

Still another such problem, the minor containment problem, is to find a fixed graph as a minor of a given graph. A minor or subcontraction of a graph is any graph obtained by taking a subgraph and contracting some (or no) edges. Many graph properties are hereditary for minors, which means that a graph has a property if and only if all minors have it too. For example, Wagner's Theorem states:

- A graph is planar if it contains as a minor neither the complete bipartite graph $K_{3,3}$ nor the complete graph K_5.

A similar problem, the subdivision containment problem, is to find a fixed graph as a subdivision of a given graph. A subdivision or homeomorphism of a graph is any graph obtained by subdividing some (or no) edges. Subdivision containment is related to graph properties such as planarity. For example, Kuratowski's Theorem states:

- A graph is planar if it contains as a subdivision neither the complete bipartite graph $K_{3,3}$ nor the complete graph K_5.

Another problem in subdivision containment is the Kelmans–Seymour conjecture:

- Every 5-vertex-connected graph that is not planar contains a subdivision of the 5-vertex complete graph K_5.

Another class of problems has to do with the extent to which various species and generalizations of graphs are determined by their point-deleted subgraphs. For example:

- The reconstruction conjecture.

Graph Coloring

Many problems and theorems in graph theory have to do with various ways of coloring graphs. Typically, one is interested in coloring a graph so that no two adjacent vertices have the same color, or with other similar restrictions. One may also consider coloring edges (possibly so that no two coincident edges are the same color), or other variations. Among the famous results and conjectures concerning graph coloring are the following:

- Four-color theorem.
- Strong perfect graph theorem.
- Erdős–Faber–Lovász conjecture (unsolved).
- Total coloring conjecture, also called Behzad's conjecture (unsolved).
- List coloring conjecture (unsolved).
- Hadwiger conjecture (graph theory) (unsolved).

Subsumption and Unification

Constraint modeling theories concern families of directed graphs related by a partial order. In these applications, graphs are ordered by specificity, meaning that more constrained graphs—which are more specific and thus contain a greater amount of information—are subsumed by those that are more general. Operations between graphs include evaluating the direction of a subsumption relationship between two graphs, if any, and computing graph unification. The unification of two argument graphs is defined as the most general graph (or the computation thereof) that is consistent with (i.e. contains all of the information in) the inputs, if such a graph exists; efficient unification algorithms are known.

For constraint frameworks which are strictly compositional, graph unification is the sufficient satisfiability and combination function. Well-known applications include automatic theorem proving and modeling the elaboration of linguistic structure.

Route Problems

- Hamiltonian path problem.
- Minimum spanning tree.
- Route inspection problem (also called the "Chinese postman problem").
- Seven bridges of Königsberg.
- Shortest path problem.
- Steiner tree.
- Three-cottage problem.
- Traveling salesman problem (NP-hard).

Network Flow

There are numerous problems arising especially from applications that have to do with various notions of flows in networks, for example:

- Max flow min cut theorem.

Visibility Problems

- Museum guard problem.

Covering Problems

Covering problems in graphs may refer to various set cover problems on subsets of vertices/subgraphs.

- Dominating set problem is the special case of set cover problem where sets are the closed neighborhoods.
- Vertex cover problem is the special case of Set cover problem where sets to cover are every edges.
- The original set cover problem, also called hitting set, can be described as a vertex cover in a hypergraph.

Decomposition Problems

Decomposition, defined as partitioning the edge set of a graph (with as many vertices as necessary accompanying the edges of each part of the partition), has a wide variety of question. Often, it is required to decompose a graph into subgraphs isomorphic to a fixed graph; for instance, decomposing a complete graph into Hamiltonian cycles. Other problems specify a family of graphs into which a given graph should be decomposed, for instance, a family of cycles, or decomposing a complete graph K_n into n – 1 specified trees having, respectively, 1, 2, 3,..., n – 1 edges.

Some specific decomposition problems that have been studied include:

- Arboricity, a decomposition into as few forests as possible.
- Cycle double cover, a decomposition into a collection of cycles covering each edge exactly twice.
- Edge coloring, a decomposition into as few matchings as possible.
- Graph factorization, a decomposition of a regular graph into regular subgraphs of given degrees.

Graph Classes

Many problems involve characterizing the members of various classes of graphs. Some examples of such questions are below:

- Enumerating the members of a class.
- Characterizing a class in terms of forbidden substructures.
- Ascertaining relationships among classes (e.g. does one property of graphs imply another).
- Finding efficient algorithms to decide membership in a class.
- Finding representations for members of a class.

Geometric Graph Theory

Geometric graph theory in the broader sense is a large and amorphous subfield of graph theory, concerned with graphs defined by geometric means. In a stricter sense, geometric graph theory studies combinatorial and geometric properties of geometric graphs, meaning graphs drawn in the Euclidean plane with possibly intersecting straight-line edges, and topological graphs, where the edges are allowed to be arbitrary continuous curves connecting the vertices, thus it is "the theory of geometric and topological graphs".

Different Types of Geometric Graphs

A planar straight line graph is a graph in which the vertices are embedded as points in the Euclidean plane, and the edges are embedded as non-crossing line segments. Fáry's theorem states that any planar graph may be represented as a planar straight line graph. A triangulation is a planar straight line graph to which no more edges may be added, so called because every face is necessarily a triangle; a special case of this is the Delaunay triangulation, a graph defined from a set of points in the plane by connecting two points with an edge whenever there exists a circle containing only those two points.

The 1-skeleton of a polyhedron or polytope is the set of vertices and edges of the polytope. The skeleton of any convex polyhedron is a planar graph, and the skeleton of any k-dimensional convex polytope is a k-connected graph. Conversely, Steinitz's theorem states that any 3-connected planar graph is the skeleton of a convex polyhedron; for this reason, this class of graphs is also known as the polyhedral graphs.

A Euclidean graph is a graph in which the vertices represent points in the plane, and the edges are assigned lengths equal to the Euclidean distance between those points. The Euclidean minimum spanning tree is the minimum spanning tree of a Euclidean complete graph. It is also possible to define graphs by conditions on the distances; in particular, a unit distance graph is formed by connecting pairs of points that are a unit distance apart in the plane. The Hadwiger–Nelson problem concerns the chromatic number of these graphs.

An intersection graph is a graph in which each vertex is associated with a set and in which vertices are connected by edges whenever the corresponding sets have a nonempty intersection. When the sets are geometric objects, the result is a geometric graph. For instance, the intersection graph of line segments in one dimension is an interval graph; the intersection graph of unit disks in the plane is a unit disk graph. The Circle packing theorem states that the intersection graphs of non-crossing circles are exactly the planar graphs. Scheinerman's conjecture states that every planar graph can be represented as the intersection graph of line segments in the plane.

A Levi graph of a family of points and lines has a vertex for each of these objects and an edge for every incident point-line pair. The Levi graphs of projective configurations lead to many important symmetric graphs and cages.

The visibility graph of a closed polygon connects each pair of vertices by an edge whenever the line segment connecting the vertices lies entirely in the polygon. It is not known how to test efficiently whether an undirected graph can be represented as a visibility graph.

A partial cube is a graph for which the vertices can be associated with the vertices of a hypercube, in such a way that distance in the graph equals Hamming distance between the corresponding hypercube vertices. Many important families of combinatorial structures, such as the acyclic orientations of a graph or the adjacencies between regions in a hyperplane arrangement, can be represented as partial cube graphs. An important special case of a partial cube is the skeleton of the permutohedron, a graph in which vertices represent permutations of a set of ordered objects and edges represent swaps of objects adjacent in the order. Several other important classes of graphs including median graphs have related definitions involving metric embeddings.

A flip graph is a graph formed from the triangulations of a point set, in which each vertex represents a triangulation and two triangulations are connected by an edge if they differ by the replacement of one edge for another. It is also possible to define related flip graphs for partitions into quadrilaterals or pseudotriangles, and for higher-dimensional triangulations. The flip graph of triangulations of a convex polygon forms the skeleton of the associahedron or Stasheff polytope. The flip graph of the regular triangulations of a point set (projections of higher-dimensional convex hulls) can also be represented as a skeleton, of the so-called secondary polytope.

ORDER THEORY

Order theory is a branch of mathematics which investigates the intuitive notion of order using binary relations. It provides a formal framework for describing statements such as "this is less than that" or "this precedes that".

Partially Ordered Sets

Orders are special binary relations. Suppose that P is a set and that ≤ is a relation on P. Then ≤ is a partial order if it is reflexive, antisymmetric, and transitive, i.e., for all a, b and c in P, we have that:

- a ≤ a (reflexivity)
- if a ≤ b and b ≤ a then a = b (antisymmetry)
- if a ≤ b and b ≤ c then a ≤ c (transitivity).

A set with a partial order on it is called a partially ordered set, poset, or just an ordered set if the intended meaning is clear. By checking these properties, one immediately sees that the well-known orders on natural numbers, integers, rational numbers and reals are all orders in the above sense. However, these examples have the additional property of being connex, i.e., for all a and b in P, we have that:

a ≤ b or b ≤ a (connexity).

A connex partial order is called a total order. These orders can also be called linear orders or chains. While many classical orders are linear, the subset order on sets provides an example where this is not the case. Another example is given by the divisibility (or "is-a-factor-of") relation |. For two natural numbers n and m, we write n|m if n divides m without remainder. One easily sees that this yields a partial order. The identity relation = on any set is also a partial order in which every two distinct elements are incomparable. It is also the only relation that is both a partial order and an equivalence relation. Many advanced properties of posets are interesting mainly for non-linear orders.

Visualizing a Poset

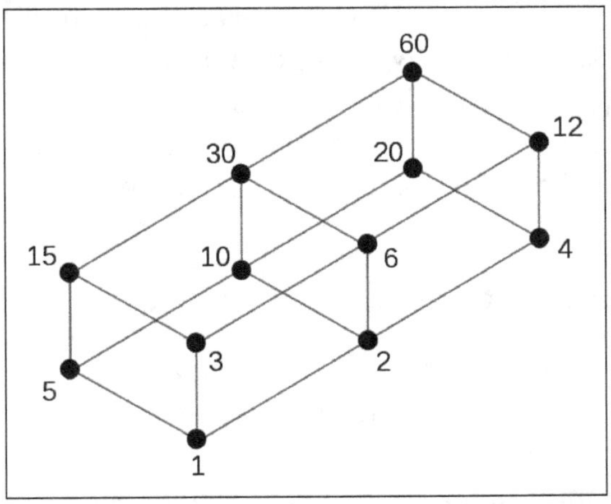

Hasse diagram of the set of all divisors of 60, partially ordered by divisibility.

Hasse diagrams can visually represent the elements and relations of a partial ordering. These are graph drawings where the vertices are the elements of the poset and the ordering relation is indicated by both the edges and the relative positioning of the vertices. Orders are drawn bottom-up: if an element x is smaller than (precedes) y then there exists a path from x to y that is directed upwards. It is often necessary for the edges connecting elements to cross each other, but elements must never be located within an edge. An instructive exercise is to draw the Hasse diagram for the set of natural numbers that are smaller than or equal to 13, ordered by the divides relation.

Even some infinite sets can be diagrammed by superimposing an ellipsis on a finite sub-order. This works well for the natural numbers, but it fails for the reals, where there is no immediate successor above 0; however, quite often one can obtain an intuition related to diagrams of a similar kind.

Special Elements within an Order

In a partially ordered set there may be some elements that play a special role. The most basic example is given by the least element of a poset. For example, 1 is the least element of the positive integers and the empty set is the least set under the subset order. Formally, an element m is a least element if:

$m \leq a$, for all elements a of the order.

The notation 0 is frequently found for the least element, even when no numbers are concerned. However, in orders on sets of numbers, this notation might be inappropriate or ambiguous, since the number 0 is not always least. An example is given by the above divisibility order |, where 1 is the least element since it divides all other numbers. In contrast, 0 is the number that is divided by all other numbers. Hence it is the greatest element of the order. Other frequent terms for the least and greatest elements is bottom and top or zero and unit.

Least and greatest elements may fail to exist, as the example of the real numbers shows. But if they exist, they are always unique. In contrast, consider the divisibility relation | on the set {2,3,4,5,6}. Although this set has neither top nor bottom, the elements 2, 3, and 5 have no elements below them, while 4, 5 and 6 have none above. Such elements are called minimal and maximal, respectively. Formally, an element m is minimal if:

$a \leq m$ implies $a = m$, for all elements a of the order.

Exchanging ≤ with ≥ yields the definition of maximality. As the example shows, there can be many maximal elements and some elements may be both maximal and minimal (e.g. 5 above). However, if there is a least element, then it is the only minimal element of the order. Again, in infinite posets maximal elements do not always exist - the set of all finite subsets of a given infinite set, ordered by subset inclusion, provides one of many counterexamples. An important tool to ensure the existence of maximal elements under certain conditions is Zorn's Lemma.

Subsets of partially ordered sets inherit the order. We already applied this by considering the subset {2, 3, 4, 5, 6} of the natural numbers with the induced divisibility ordering. Now there are also elements of a poset that are special with respect to some subset of the order. This leads to the definition of upper bounds. Given a subset S of some poset P, an upper bound of S is an element b of P that is above all elements of S. Formally, this means that,

$s \leq b$, for all s in S.

Lower bounds again are defined by inverting the order. For example, -5 is a lower bound of the natural numbers as a subset of the integers. Given a set of sets, an upper bound for these sets under the subset ordering is given by their union. In fact, this upper bound is quite special: it is the smallest set that contains all of the sets. Hence, we have found the least upper bound of a set of sets. This concept is also called supremum or join, and for a set S one writes sup(S) or $S\vee$ for its least upper bound. Conversely, the greatest lower bound is known as infimum or meet and denoted inf(S) or $S\wedge$. These concepts play an important role in many applications of order theory. For two elements x and y, one also writes $x \vee y$ and $x \wedge y$ for sup({x,y}) and inf({x,y}), respectively.

For example, 1 is the infimum of the positive integers as a subset of integers. For another example, consider again the relation | on natural numbers. The least upper bound of two numbers is the smallest number that is divided by both of them, i.e. the least common multiple of the numbers. Greatest lower bounds in turn are given by the greatest common divisor.

Duality

In the previous definitions, we often noted that a concept can be defined by just inverting the ordering in a former definition. This is the case for "least" and "greatest", for "minimal" and "maximal", for "upper bound" and "lower bound", and so on. This is a general situation in order theory: A given order can be inverted by just exchanging its direction, pictorially flipping the Hasse diagram top-down. This yields the so-called dual, inverse, or opposite order.

Every order theoretic definition has its dual: it is the notion one obtains by applying the definition to the inverse order. Since all concepts are symmetric, this operation preserves the theorems of partial orders. For a given mathematical result, one can just invert the order and replace all definitions by their duals and one obtains another valid theorem. This is important and useful, since one obtains two theorems for the price of one.

Constructing New Orders

There are many ways to construct orders out of given orders. The dual order is one example. Another important construction is the cartesian product of two partially ordered sets, taken together with the product order on pairs of elements. The ordering is defined by (a, x) ≤ (b, y) if (and only if) a ≤ b and x ≤ y. (Notice carefully that there are three distinct meanings for the relation symbol ≤ in this definition.) The disjoint union of two posets is another typical example of order construction, where the order is just the (disjoint) union of the original orders.

Every partial order ≤ gives rise to a so-called strict order <, by defining a < b if a ≤ b and not b ≤ a. This transformation can be inverted by setting a ≤ b if a < b or a = b. The two concepts are equivalent although in some circumstances one can be more convenient to work with than the other.

Functions between Orders

It is reasonable to consider functions between partially ordered sets having certain additional properties that are related to the ordering relations of the two sets. The most fundamental condition that occurs in this context is monotonicity. A function f from a poset P to a poset Q is monotone, or order-preserving, if a ≤ b in P implies f(a) ≤ f(b) in Q (Noting that, strictly, the two relations here are different since they apply to different sets.). The converse of this implication leads to functions that are order-reflecting, i.e. functions f as above for which f(a) ≤ f(b) implies a ≤ b. On the other hand, a function may also be order-reversing or antitone, if a ≤ b implies f(a) ≥ f(b).

An order-embedding is a function f between orders that is both order-preserving and order-reflecting. Examples for these definitions are found easily. For instance, the function that maps a natural number to its successor is clearly monotone with respect to the natural order. Any function from a discrete order, i.e. from a set ordered by the identity order "=", is also monotone. Mapping each natural number to the corresponding real number gives an example for an order embedding. The set complement on a powerset is an example of an antitone function.

An important question is when two orders are "essentially equal", i.e. when they are the same up to renaming of elements. Order isomorphisms are functions that define such a renaming. An order-isomorphism is a monotone bijective function that has a monotone inverse. This is equivalent to being a surjective order-embedding. Hence, the image f(P) of an order-embedding is always isomorphic to P, which justifies the term "embedding".

A more elaborate type of functions is given by so-called Galois connections. Monotone Galois connections can be viewed as a generalization of order-isomorphisms, since they constitute of a pair of two functions in converse directions, which are "not quite" inverse to each other, but that still have close relationships.

Another special type of self-maps on a poset are closure operators, which are not only monotonic, but also idempotent, i.e. f(x) = f(f(x)), and extensive (or inflationary), i.e. x ≤ f(x). These have many applications in all kinds of "closures" that appear in mathematics.

Besides being compatible with the mere order relations, functions between posets may also behave well with respect to special elements and constructions. For example, when talking about posets with least element, it may seem reasonable to consider only monotonic functions that preserve this element, i.e. which map least elements to least elements. If binary infima ∧ exist, then a reasonable property might be to require that $f(x \wedge y) = f(x) \wedge f(y)$, for all x and y. All of these properties, and indeed many more, may be compiled under the label of limit-preserving functions.

Finally, one can invert the view, switching from functions of orders to orders of functions. Indeed, the functions between two posets P and Q can be ordered via the pointwise order. For two functions f and g, we have $f \leq g$ if $f(x) \leq g(x)$ for all elements x of P. This occurs for example in domain theory, where function spaces play an important role.

Special Types of Orders

Many of the structures that are studied in order theory employ order relations with further properties. In fact, even some relations that are not partial orders are of special interest. Mainly the concept of a preorder has to be mentioned. A preorder is a relation that is reflexive and transitive, but not necessarily antisymmetric. Each preorder induces an equivalence relation between elements, where a is equivalent to b, if $a \leq b$ and $b \leq a$. Preorders can be turned into orders by identifying all elements that are equivalent with respect to this relation.

Several types of orders can be defined from numerical data on the items of the order: a total order results from attaching distinct real numbers to each item and using the numerical comparisons to order the items; instead, if distinct items are allowed to have equal numerical scores, one obtains a strict weak ordering. Requiring two scores to be separated by a fixed threshold before they may be compared leads to the concept of a semiorder, while allowing the threshold to vary on a per-item basis produces an interval order.

An additional simple but useful property leads to so-called well-founded, for which all non-empty subsets have a minimal element. Generalizing well-orders from linear to partial orders, a set is well partially ordered if all its non-empty subsets have a finite number of minimal elements.

Many other types of orders arise when the existence of infima and suprema of certain sets is guaranteed. Focusing on this aspect, usually referred to as completeness of orders, one obtains:

- Bounded posets, i.e. posets with a least and greatest element (which are just the supremum and infimum of the empty subset),
- Lattices, in which every non-empty finite set has a supremum and infimum,
- Complete lattices, where every set has a supremum and infimum, and

- Directed complete partial orders (dcpos), that guarantee the existence of suprema of all directed subsets and that are studied in domain theory.

Partial orders with complements, or poc sets, are posets with a unique bottom element 0, as well as an order-reversing involution * such that $a \leq a^* \Rightarrow a = 0$.

However, one can go even further: if all finite non-empty infima exist, then ∧ can be viewed as a total binary operation in the sense of universal algebra. Hence, in a lattice, two operations ∧ and ∨ are available, and one can define new properties by giving identities, such as,

$$x \wedge (y \vee z) = (x \wedge y) \vee (x \wedge z), \text{ for all } x, y, \text{ and } z.$$

This condition is called distributivity and gives rise to distributive lattices. There are some other important distributivity laws Some additional order structures that are often specified via algebraic operations and defining identities are:

- Heyting algebras.
- Boolean algebras.

which both introduce a new operation ~ called negation. Both structures play a role in mathematical logic and especially Boolean algebras have major applications in computer science. Finally, various structures in mathematics combine orders with even more algebraic operations, as in the case of quantales, that allow for the definition of an addition operation.

Many other important properties of posets exist. For example, a poset is locally finite if every closed interval [a, b] in it is finite. Locally finite posets give rise to incidence algebras which in turn can be used to define the Euler characteristic of finite bounded posets.

Subsets of Ordered Sets

In an ordered set, one can define many types of special subsets based on the given order. A simple example are upper sets; i.e. sets that contain all elements that are above them in the order. Formally, the upper closure of a set S in a poset P is given by the set {x in P | there is some y in S with y ≤ x}. A set that is equal to its upper closure is called an upper set. Lower sets are defined dually.

More complicated lower subsets are ideals, which have the additional property that each two of their elements have an upper bound within the ideal. Their duals are given by filters. A related concept is that of a directed subset, which like an ideal contains upper bounds of finite subsets, but does not have to be a lower set. Furthermore, it is often generalized to preordered sets.

A subset which is - as a sub-poset - linearly ordered, is called a chain. The opposite notion, the antichain, is a subset that contains no two comparable elements; i.e. that is a discrete order.

Related Mathematical Areas

Although most mathematical areas use orders in one or the other way, there are also a few theories that have relationships which go far beyond mere application.

Universal Algebra

The methods and formalisms of universal algebra are an important tool for many order theoretic considerations. Beside formalizing orders in terms of algebraic structures that satisfy certain identities, one can also establish other connections to algebra. An example is given by the correspondence between Boolean algebras and Boolean rings. Other issues are concerned with the existence of free constructions, such as free lattices based on a given set of generators. Furthermore, closure operators are important in the study of universal algebra.

Topology

In topology, orders play a very prominent role. In fact, the collection of open sets provides a classical example of a complete lattice, more precisely a complete Heyting algebra (or "frame" or "locale"). Filters and nets are notions closely related to order theory and the closure operator of sets can be used to define a topology. Beyond these relations, topology can be looked at solely in terms of the open set lattices, which leads to the study of pointless topology. Furthermore, a natural preorder of elements of the underlying set of a topology is given by the so-called specialization order, that is actually a partial order if the topology is T_0.

Conversely, in order theory, one often makes use of topological results. There are various ways to define subsets of an order which can be considered as open sets of a topology. Considering topologies on a poset (X, ≤) that in turn induce ≤ as their specialization order, the finest such topology is the Alexandrov topology, given by taking all upper sets as opens. Conversely, the coarsest topology that induces the specialization order is the upper topology, having the complements of principal ideals (i.e. sets of the form {y in X | y ≤ x} for some x) as a subbase. Additionally, a topology with specialization order ≤ may be order consistent, meaning that their open sets are "inaccessible by directed suprema" (with respect to ≤). The finest order consistent topology is the Scott topology, which is coarser than the Alexandrov topology. A third important topology in this spirit is the Lawson topology. There are close connections between these topologies and the concepts of order theory. For example, a function preserves directed suprema if and only if it is continuous with respect to the Scott topology (for this reason this order theoretic property is also called Scott-continuity).

Category Theory

The visualization of orders with Hasse diagrams has a straightforward generalization: Instead of displaying lesser elements below greater ones, the direction of the order can also be depicted by giving directions to the edges of a graph. In this way, each order is seen to be equivalent to a directed acyclic graph, where the nodes are the elements of the poset and there is a directed path from a to b if and only if $a \leq b$. Dropping the requirement of being acyclic, one can also obtain all preorders.

When equipped with all transitive edges, these graphs in turn are just special categories, where elements are objects and each set of morphisms between two elements is at most singleton. Functions between orders become functors between categories. Many ideas of order theory are just concepts of category theory in small. For example, an infimum is just a categorical product. More generally, one can capture infima and suprema under the abstract notion of a categorical limit (or colimit, respectively). Another place where categorical ideas occur is the concept of a (monotone) Galois connection, which is just the same as a pair of adjoint functors.

But category theory also has its impact on order theory on a larger scale. Classes of posets with appropriate functions form interesting categories. Often one can also state constructions of orders, like the product order, in terms of categories. Further insights result when categories of orders are found categorically equivalent to other categories, for example of topological spaces. This line of research leads to various representation theorems, often collected under the label of Stone duality.

DISCRETE MORSE THEORY

Discrete Morse theory is a combinatorial adaptation of Morse theory developed by Robin Forman. The theory has various practical applications in diverse fields of applied mathematics and computer science, such as configuration spaces, homology computation, denoising, mesh compression, and topological data analysis.

Notation Regarding CW Complexes

Let $\mathcal{X}$ be a CW complex. Define the incidence function $\kappa : \mathcal{X} \times \mathcal{X} \to \mathbb{Z}$ in the following way: given two cells σ and τ in let $\kappa(\sigma, \tau)$ be the degree of the attaching map from the boundary of σ to τ. The boundary operator ∂ on $\mathcal{X}$ is defined by,

$$\partial(\sigma) = \sum_{\tau \in \mathcal{X}} \kappa(\sigma, \tau)\tau$$

It is a defining property of boundary operators that $\partial^{\circ}\partial \equiv 0$. In more axiomatic definitions one can find the requirement that $\forall \sigma, \tau' \in \mathcal{X}$,

$$\sum_{\tau \in \mathcal{X}} \kappa(\sigma, \tau)\kappa(\tau, \tau') = 0$$

which is a corollary of the above definition of the boundary operator and the requirement that $\partial \circ \partial \equiv 0$.

Discrete Morse Functions

A real-valued function $\mu:\mathcal{X}\to\mathbb{R}$ is a discrete Morse function if it satisfies the following two properties:

- For any cell $\sigma\in\mathcal{X}$, the number of cells $\tau\in\mathcal{X}$ in the boundary of σ which satisfy $\mu(\sigma)\le\mu(\tau)$ is at most one.
- For any cell $\sigma\in\mathcal{X}$, the number of cells $\hat{o}\in\mathcal{X}$ containing σ in their boundary which satisfy $\mu(\sigma)\ge\mu(\tau)$ is at most one.

It can be shown that the cardinalities in the two conditions cannot both be one simultaneously for a fixed cell σ, provided that $\mathcal{X}$ is a regular CW complex. In this case, each cell $\sigma\in\mathcal{X}$ can be paired with at most one exceptional cell $\tau\in\mathcal{X}$: Either a boundary cell with larger $\imath$ value, or a co-boundary cell with smaller $\imath$ value. The cells which have no pairs, i.e., whose function values are strictly higher than their boundary cells and strictly lower than their co-boundary cells are called critical cells. Thus, a discrete Morse function partitions the CW complex into three distinct cell collections: $\mathcal{X}=\mathcal{A}\sqcup\mathcal{K}\sqcup\mathcal{Q}$, where:

- $\mathcal{A}$ denotes the critical cells which are unpaired,
- $\mathcal{K}$ denotes cells which are paired with boundary cells,
- $\mathcal{Q}$ denotes cells which are paired with co-boundary cells.

By construction, there is a bijection of sets between $k-$dimensional cells in $\mathcal{K}$ and the $(k-1)-$ dimensional cells in $\mathcal{Q}$, which can be denoted by $p^k:\mathcal{K}^k\to\mathcal{Q}^{k-1}$ for each natural number k. It is an additional technical requirement that for each $K\in\mathcal{K}^k$,, the degree of the attaching map from the boundary of K to its paired cell $p^k(K)\in\mathcal{Q}$ is a unit in the underlying ring of $\mathcal{X}$. For instance, over the integers $\mathbb{Z}$, the only allowed values are ± 1. This technical requirement is guaranteed, for instance, when one assumes that $\mathcal{X}$ is a regular CW complex over $\mathbb{Z}$.

The fundamental result of discrete Morse theory establishes that the CW complex $\mathcal{X}$ is isomorphic on the level of homology to a new complex $\mathcal{A}$ consisting of only the critical cells. The paired cells in $\mathcal{K}$ and describe gradient paths between adjacent critical cells which can be used to obtain the boundary operator on $\mathcal{A}$.

The Morse Complex

A gradient path is a sequence of paired cells,

$$\rho=(Q_1,K_1,Q_2,K_2,\ldots,Q_M,K_M)$$

satisfying $Q_m=p(K_m)$ and $\kappa(K_m,Q_{m+1})\ne 0$. The index of this gradient path is defined to be the integer,

$$\nu(\rho)=\frac{\prod_{m=1}^{M-1}-\kappa(K_m,Q_{m+1})}{\prod_{m=1}^{M-1}\kappa(K_m,Q_m)}.$$

The division here makes sense because the incidence between paired cells must be ±1. Note that by construction, the values of the discrete Morse function μ must decrease across $\tilde{n}$. The path $\tilde{n}$ is said to connect two critical cells $A, A' \in \mathcal{A}$ if,

$$\kappa(A, Q_1) \neq 0 \neq \kappa(K_M, A').$$

This relationship may be expressed as $A \xrightarrow{\rho} A'$. The multiplicity of this connection is defined to be the integer $m(\rho) = \kappa(A, Q_1) \cdot \nu(\rho) \cdot \kappa(K_M, A')$. Finally, the Morse boundary operator on the critical cells $\mathcal{A}$ is defined by,

$$\Delta(A) = \kappa(A, A') + \sum_{A \xrightarrow{\rho} A'} m(\rho) A'$$

where the sum is taken over all gradient path connections from A to A′.

Many of the familiar results from continuous Morse theory apply in the discrete setting.

The Morse Inequalities

Let $\mathcal{A}$ be a Morse complex associated to the CW complex $\mathcal{X}$. The number $m_q = |\mathcal{A}_q|$ of $q-$cells in $\mathcal{A}$ is called the qth Morse number. Let β_q denote the qth Betti number of $\mathcal{X}$. Then, for any $N > 0$, the following inequalities hold,

$m_N \geq \beta_N$, and

$$m_N - m_{N-1} + \ldots \pm m_0 \geq \beta_N - \beta_{N-1} + \ldots \pm \beta_0$$

Moreover, the Euler characteristic $\chi(\mathcal{X})$ of $\mathcal{X}$ satisfies

$$\chi(\mathcal{X}) = m_0 - m_1 + \ldots \pm m_{\dim \mathcal{X}}$$

Discrete Morse Homology and Homotopy Type

Let $\mathcal{X}$ be a regular CW complex with boundary operator ∂ and a discrete Morse function $\mu : \mathcal{X} \to \mathbb{R}$. Let $\mathcal{A}$ be the associated Morse complex with Morse boundary operator Δ. Then, there is an isomorphism of homology groups:

$$H_*(\mathcal{X}, \partial) \simeq H_*(\mathcal{A}, \Delta),$$

and similarly for the homotopy groups.

COMBINATORIAL DESIGN THEORY

Combinatorial design theory involves the study of finite objects satisfying certain balance and symmetry conditions.

Regular Graphs

A d-regular graph is a graph where all of the vertex degrees are d. Equivalently, it is a symmetric 0-1 matrix with zeros on the main diagonal, whose rows and columns sum to d. These objects come up in innumerable settings, and they have been well studied. In particular, a long line of work involving Read, Mckay, Wormald, Bollobas, Bender and Canfield and many others established asymptotic fomulae for the number of d-regular graphs on n vertices when $d = o(\sqrt{n})$.

For a constant d, there is a simple algorithm, called the configuration model, for sampling d-regular graphs uniformly at random. Start with nd "half edges" (v, i) where v $\in V$ and $1 \le i \le d$, and choose a perfect matching uniformly at random on this set. This induces a regular (possibly non-simple) graph G that may include loops and multiple edges. The chance that G is simple turns out to be dependent only on d and not on n, so when d is a constant, there is a constant chance that we get a simple graph. Conditioned on G being simple, it is uniformly distributed on the set of d-regular graphs.

Perfect Matchings in Hypergraphs

A d-uniform hypergraph is a pair $H = \langle V,F \rangle$ such that $F \subseteq \binom{V}{d}$ perfect matching in H is a collection M of hyperedges of H such that each vertex belongs to a single edge of M.

The adjacency matrix of a graph G is the n × n 0-1 matrix A_G defined by $A_G(i, j) = 1$ iff $\{i, j\} \in E(G)$. Just as a perfect matching in G is equivalent to a symmetric permutation matrix contained in A_G, perfect matchings in hypergraphs also have a matrix representation, but it involves d-dimensional matrices.

The adjacency matrix of a d-uniform order-n hypergraph H is the $[n]^{d}$ matrix A_H such that $AH(i_1,..., i_d) = 1$ iff $\{i_1,..., i_d\} \in F(H)$. A perfect matching H is equivalent to a 0-1 [n] d matrix X contained in AH such that:

- Each hyperplane in X contains a unique one:

$$\sum_{i_1,\dots,i_{k-1},i_{k+1},\dots,i_d} X(i_1,\dots,i_d) = 1 \qquad \text{for all } 1 \le k \le d \text{ and } 1 \le i_k \le n.$$

- X is totally symmetric:

$$X(i_1, \dots, i_d) = X(i_{\sigma 1}, \dots, i_{\sigma d}) \qquad \text{for every } \sigma \in \mathbb{S}_d.$$

The total number of d-uniform perfect matchings on n vertices is easy to compute. Clearly, n must be a multiple of d, and when this happens, the number of perfect matchings is $\binom{n}{d,\dots,d}$. However, the number of perfect matchings in a given hypergraph H is #P hard to compute for $d \ge 2$. For those unversed in complexity theory, this means that if P 6= NP, then there is no efficient algorithm that computes it. When $d \ge 3$ there is (probably) not even an efficient algorithm to check whether H has a single perfect matching, in contrast to the graph case, where such algorithms are known.

A recent breakthrough in the study of these objects was the computation of the threshold for the appearance of a perfect matching in a random hypergraph where every hyperedge appears with probability p. It was shown that, as in the graph case, a perfect matching appears shortly after there are no more isolated vertices.

Latin Squares

An order-n Latin square is an $n \times n$ matrix over the symbols $[n] := \{1,\dots, n\}$ such that each row and each column is a permutation. The following is an example of a Latin square:

$$L = \begin{pmatrix} 1 & 2 & 3 & 4 \\ 2 & 3 & 4 & 1 \\ 3 & 4 & 1 & 2 \\ 4 & 1 & 2 & 3 \end{pmatrix}$$

Latin squares also have an equivalent 3d matrix representation: A Latin square is equivalent to an $n \times n \times n$ 0-1 matrix A with a single one in each line, where a line is the set of entries obtained by fixing all but one of the indices and allowing that index to vary over [n]. The equivalence is given via $L(i, j) = k \Leftrightarrow A(i, j, k) = 1$. The Latin square L may be viewed as a "topographical map" of A. For example, the 3d representation of the Latin square above is,

$$A = \left(\begin{pmatrix} 1 & 0 & 0 & 0 \\ 0 & 0 & 0 & 1 \\ 0 & 0 & 1 & 0 \\ 0 & 1 & 0 & 0 \end{pmatrix}, \begin{pmatrix} 0 & 1 & 0 & 0 \\ 1 & 0 & 0 & 0 \\ 0 & 0 & 0 & 1 \\ 0 & 0 & 1 & 0 \end{pmatrix}, \begin{pmatrix} 0 & 0 & 1 & 0 \\ 0 & 1 & 0 & 0 \\ 1 & 0 & 0 & 0 \\ 0 & 0 & 0 & 1 \end{pmatrix}, \begin{pmatrix} 0 & 0 & 0 & 1 \\ 0 & 0 & 1 & 0 \\ 0 & 1 & 0 & 0 \\ 1 & 0 & 0 & 0 \end{pmatrix} \right)$$

One useful fact about Latin squares is that it is easy to construct them. For one thing, the multiplication table of any group is a Latin square (since $ab = c$ and $ad = c$ implies $b = d$). For another thing, we can construct the Latin square row by row and we will never get stuck. Choosing the next row in the partial Latin square reduces to choosing a perfect matching in a regular bipartite graph, and this is always possible because of Hall's marriage theorem.

This last insight yields good bounds on the number Ln of order-n Latin squares. Without going into details, permanent bounds can also bound the number of perfect matchings in a bipartite graph, and so by using these bounds to estimate the number of possibilities for each row, we can show that,

$$L_n = \left((1 + o(1)) \frac{n}{e^2} \right)^{n^2}.$$

1-Factorizations

A 1-factorization of a d-regular graph G is a partition of its edge set into d 1-factors, or perfect matchings. Equivalently, it is a proper edge-coloring of G using d colors. 1-factorizations of the complete graph K_n are equivalent to symmetric Latin squares with n on the main diagonal, via $L(i, j) = c(\{i, j\})$. We may view the relationship between Latin squares and 1-factorizations as analogous to that of Adjacency matrices of bipartite graphs (which need not be symmetric) and adjacency matrices of general graphs, which are symmetric and have zeros on the main diagonal. It is interesting to note that Latin squares are also equivalent to 1-factorizations of the complete bipartite graph $K_{n,n}$, and this fact yields a nice construction of 1-factorizations.

Assume that n is a power of 2. We will construct a 1-factorization of Kn recursively as follows. Split [n] into two sets of size $\frac{n}{2}$, color the edges between them using a Latin square and the edges inside each set recursively. This construction implies that $F(n) \geq F(n/2)^2 \cdot L_{n/2}$, which yields the lower bound,

$$F(n) \geq \left((1+o(1))\frac{n}{4e^2}\right)^{\frac{n^2}{2}}.$$

By adding perfect matchings one by one and upper bounding the number of available perfect matchings at each step using standard bounds, we can show that,

$$F(n) \geq \left((1+o(1))\frac{n}{e^2}\right)^{\frac{n^2}{2}}.$$

There is a substantial gap between the lower bound and the upper bound here.

Steiner Triple Systems

A Steiner triple system (STS) is another analog of a perfect matching for 3-uniform hypergraphs.

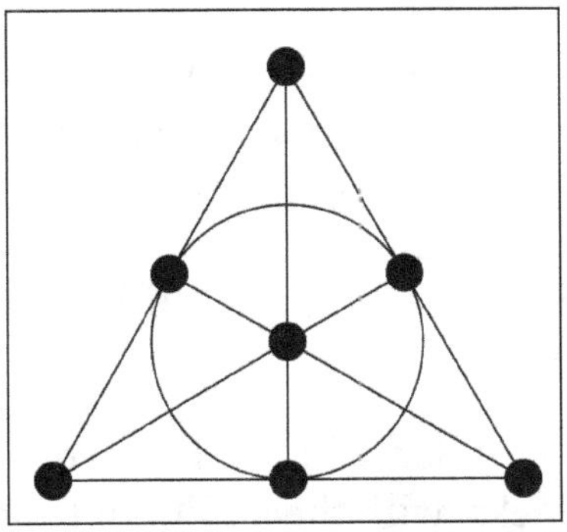

This is an example of an STS on 7 vertices. Each line represents a triple, and you may check that each pair appears in a unique triple.

- A perfect matching is a collection of pairs such that each vertex belongs to exactly one pair.
- An STS is a collection of triples such that each pair of vertices belongs to exactly one triple.

STSs are triangle decompositions of the edge set of Kn. Latin squares are equivalent to a triangle decomposition of the complete tripartite graph, and indeed it is possible to construct a Steiner triple system recursively using Latin squares. Such a construction yields the lower bound,

$$STS(n) \geq \left((1+o(1))\frac{n}{3\sqrt{3e^2}})\right)^{\frac{n^2}{6}}$$

The entropy method, is a powerful tool for estimating the number of combinatorial objects of a certain size. An entropy proof shows that,

$$STS(n) \geq \left((1+o(1))\frac{n}{e^2})\right)^{\frac{n^2}{6}}$$

Once again, there is a substantial gap.

Steiner triple systems can be represented by Latin squares. Given an STS X, define a Latin square L by L(i, j) = k if {i, j, k} ∈ X, and set L(i, i) = i for every i ∈ [n]. This Latin square is even more symmetric than the Latin squares that represent 1-factorizations: L(i, j) = k ⇔ L(k, j) = i ⇔ L(k, i) = j ⇔ and so on.

(n, q, r, λ)-Designs

The following definition is an overreaching generalization of most of the objects.

An (n, q, r, λ)-design is a collection X of q-element subsets of [n] such that every r-element subset of [n] is contained in exactly λ elements of X.

So a d-regular graph is an (n, 2, 1, d)-design, a perfect matching in a d-uniform hypergraph is an (n, d, 1, 1)-design, and an STS is an (n, 3, 2, 1)-design. This is the concept that Keevash's work deals with. He totally solved the existence and enumeration problems here. Perhaps luckily for the remaining researchers in the field of combinatorial designs, there are many interesting objects that this concept doesn't capture, such as Latin squares, 1-factorizations.

Sudoku squares: We are all familiar with Sudoku squares. They are order 9 Latin squares divided into 93 × 3 blocks, with the additional constraint that each block must contain the symbols {1,..., 9}. It is possible to define Sudoku squares of size N × N for any number $N = n^2$, and then it becomes interesting to ask how many order-N Sudoku squares there are.

Using entropy methods, which really are a very powerful tool, it is possible to show that the number of order-N Sudoku squares is at most $\left((1 + o(1))N/e^3\right)^{N^2}$, but I don't know of any good lower bounds.

Latin Transversals

Let L be an order-n Latin square. A transversal of L is a collection of n elements of L, exactly one from each row and each column and one of each symbol. Here is an example:

$$L = \begin{pmatrix} 1 & 2 & 3 & 4 & 5 \\ 2 & 3 & 4 & 5 & 1 \\ 3 & 4 & 5 & 1 & 2 \\ 4 & 5 & 1 & 2 & 3 \\ 5 & 1 & 2 & 3 & 4 \end{pmatrix}.$$

Transversals of Latin squares are well studied objects, of great importance for many problems of interest in the study of Latin squares. Here the biggest open problem remains the existence problem. It is well known that there are Latin squares of even order without any transversals, but it is conjectured that every Latin square of odd order has a transversal. This is the Ryser conjecture. Here, as far as I can see, Keevash's methods are not sufficient to solve the problem without substantial new ideas. This may be the biggest open problem remaining in the field of combinatorial designs.

A major open question about Latin transversals was answered recently, that the number of transversals in the cyclic Latin square is $\left(e^{-1/2} + o(1)\right) \cdot n!^3 / n^{n-1}$, an unbelievably precise asymptotic formula. This formula asymptotically matches the maximal possible number of transversals in a Latin square.

The n-queens Problem

This is an old problem that Euler, among others, worked on, and as a chess player it is close to my heart. For any large enough n it is possible to place n queens on an $n \times n$ chessboard so that no two attack each other. The question is: In how many ways is it possible to do this?

This question is related to the question about Latin transversals A transversal of the cyclic Latin square corresponds to a solution of the n-queens problem for "semiqueens" on the "torus". The idea is that instead of the usual square chessboard, we have a toroidal chessboard where falling off the edge of the board sends us back to the opposite edge. A semiqueen is a queen that can move diagonally only in one of the two possible directions, say from the lower left to the upper right.

References

- Davey, B. A.; Priestley, H. A. (2002). Introduction to Lattices and Order (2nd ed.). Cambridge University Press. ISBN 0-521-78451-4
- Hillman, Mathematics, Academics: whitman.edu, Retrieved 24 April, 2020
- Kepner, Jeremy; Gilbert, John (2011). Graph Algorithms in The Language of Linear Algebra. Philadelphia, Pennsylvania: SIAM. ISBN 978-0-898719-90-1
- Pach, János (2013). "The beginnings of geometric graph theory". Erdös centennial. Bolyai Soc. Math. Stud. 25. Budapest: János Bolyai Math. Soc. pp. 465–484. doi:10.1007/978-3-642-39286-3_17. MR 3203608
- Kozlov, Dmitry (2007). Combinatorial algebraic topology. Algorithms and Computation in Mathematics. 21. Berlin: Springer. ISBN 978-3540719618. MR 2361455

Theorems in Combinatorics

CHAPTER 3

Dilworth's theorem, Mirsky's theorem, Baranyai's theorem, Corners theorem, Folkman's theorem, Szemeredi's theorem, Kirchoff's theorem, Wagner's theorem, Hall's matching theorem, etc. are some of the theorems that are used within combinatorics. This chapter has been carefully written to provide an easy understanding of these theorems of combinatorics.

DILWORTH'S THEOREM

Let S be a finite partially ordered set. The size of a maximal antichain equals the size of a minimal chain cover of S. This is called the Dilworth's theorem. It is named after the mathematician Robert P. Dilworth.

The width of a finite partially ordered set S is the maximum size of an antichain in S. In other words, the width of a finite partially ordered set S is the minimum number of chains needed to cover S, i.e. the minimum number of chains such that any element of S is in at least one of the chains.

- Chain: A chain in a partially ordered set is a subset of elements which are all comparable to each other.
- Antichain: An antichain is a subset of elements, no two of which are comparable to each other.

Illustrative examples:

```
Let S be the set of divisors of 30, with divisibility as the partial
order.

Then the following chains cover S:

{1, 2, 6, 30}, {3, 15}, {5, 10}

And {2, 3, 5} is an antichain of length 3.

It is not immediately obvious, but the chain cover is minimal (though
not unique), and the antichain is maximal (though not unique).

So both definitions of width give 3 for this partially ordered set.
```

Proof of Dilworth's Theorem:

The easiest proof is by induction on the size of the set. Let d be the size of the largest antichain of S. The proof will show that S can be covered by d chains. The base case is trivial. So suppose the result has been proven for all sets smaller than S.

First, if no two elements of S are comparable, then S itself is an antichain and it can be covered by d = |S| chains each of length 1, so the result holds. Otherwise, let m be a minimal element (m <= z for all comparable z) and M be a maximal element (z <= M for all comparable z). Let T = S – {m, M}. If the largest antichain in T has size <= d – 1, then T can be covered by d – 1 chains, and so S can be covered by those plus the chain {m, M}, and the result will be proven for S. Now, suppose that the largest antichain in T has size d(it can't be larger because T is a subset of S). Call this antichain A.

The idea of the rest of the proof is to picture the Hasse diagram for S where the largest antichain consists of a horizontal strip. Take everything below the strip and everything above the strip, use induction to cover these by chains, and then link the chains together by connecting them across the strip.

That is, construct the two sets:

$$S^+ = \{x \text{ belongs to } S : x >= a \text{ for some } a \text{ belongs to } A\}$$

$$S^- = \{x \text{ belongs to } S : x <= a \text{ for some } a \text{ belongs to } A\}$$

Then S^+ U S^- must be all of S, because if it weren't then A would not be a maximal antichain in S. And S^+ U S^- = A, because if x is in the intersection, then a <= x <= b for some elements a, b belongs to A, so a and b are comparable by transitivity, so the only possibility is that a = b and they both equal x.

Since m and M are not in A, it must be the case that and m does not belong to S^+, and m does not belong to S^- so both sets), and are strictly smaller than S. The inductive hypothesis applies to both S^- and S^+, so they are both covered by d chains, each of which must contain exactly one element of A. Call them C_a^- and C_a^+. Now we can stitch together these covers to get a cover of all of S, by the chains C_a^- U {a} U C_a^+. This cover has d chains, so the result follows by induction.

MIRSKY'S THEOREM

Mirsky's theorem relates the size of an antichain cover and a chain in a poset. The definitions we have seen so far are sufficient to express the formal statement of Mirsky's theorem in Coq.

Theorem: Dual_ Dilworth: ∀ (P: FPO U), Dual_ Dilworth _statement P.

where, Dual_ Dilworth _statement is defined as,

Dual_Dilworth_statement:= fun (P: FPO U) ⇒ ∀ (m n: nat), (Is_height P m) → (∃ cover: Ensemble (Ensemble U), (Is_a_smallest_antichain_cover P cover) /\ (cardinal _ cover n)) → m=n.

It states that in any poset the maximum size of a chain is equal to the minimum number of antichains in any antichain cover. In other words, if c(P) represents the size of a smallest antichain cover of P, then height(P) = c(P).

Proof: The equality will follow if one can prove,

- Size of a chain ≤ Size of an antichain cover.
- There is an antichain cover of size equal to height(P).

Any chain shares at most one element with each antichain from an antichain cover. Moreover, every element of the chain must be covered by some antichain from the antichain cover. Hence, the size of any chain is smaller than or equal to the size of any antichain cover. We will prove proof 2 using strong induction on the size of the largest chain of P. Let m be the size of the largest chain in P, i.e, m = height(P).

Induction hypothesis: For all posets P′ of height at most m – 1, there exists an antichain cover of size equal to height(P′).

Induction Step: Let M denote the set of all maximal elements of P, i.e, M = maximal(P). Observe that M is a non-empty antichain and shares an element with every largest chain of P. Consider now the partially ordered set (P – M, ≤). The length of the largest chain in P – M is at most m – 1. On the other hand, if the length of the largest chain in P – M is less than m – 1, M must contain two or more elements that are members of the same chain, which is a contradiction. Hence, we conclude that the length of largest chain in P – M is m – 1. Using induction hypothesis there we get an antichain cover A_C of size m – 1 for P – M. Thus, we get an antichain cover $A_C \cup \{M\}$ of size m for P.

BARANYAI'S THEOREM

In combinatorial mathematics, Baranyai's theorem (proved by and named after Zsolt Baranyai) deals with the decompositions of complete hypergraphs.

The statement of the result is that if $2 \leq r < k$ are natural numbers and r divides k, then the complete hypergraph K_r^k decomposes into 1-factors. K_r^k is a hypergraph with k vertices, in which every subset of r vertices forms a hyperedge; a 1-factor of this hypergraph is a set of hyperedges that touches each vertex exactly once, or equivalently a partition of the vertices into subsets of size r. Thus, the theorem states that the k vertices of the hypergraph may be partitioned into subsets of r vertices in $\binom{k}{r}\frac{r}{k}=\binom{k-1}{r-1}$ different ways, in such a way that each r-element subset appears in exactly one of the partitions.

The case r = 2:

In the special case $r=2$, we have a complete graph K_n on n vertices, and we wish to color the edges with $\binom{n}{2}\frac{2}{n}=n-1$ colors so that the edges of each color form a perfect matching. Baranyai's theorem says that we can do this whenever n is even.

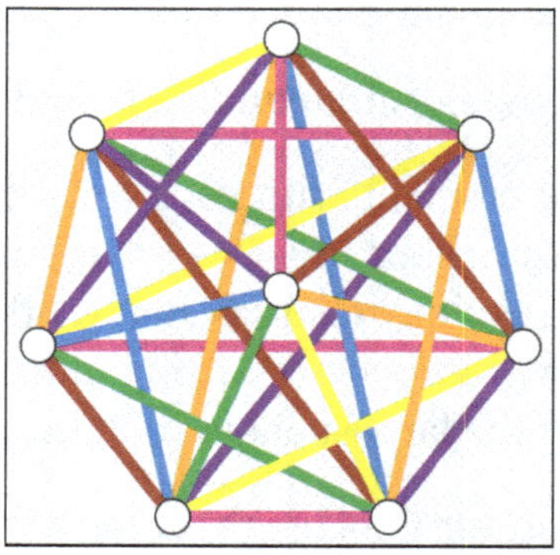

A partition of a complete graph on 8 vertices into 7 colors (perfect matchings), the case r = 2 of Baranyai's theorem.

CORNERS THEOREM

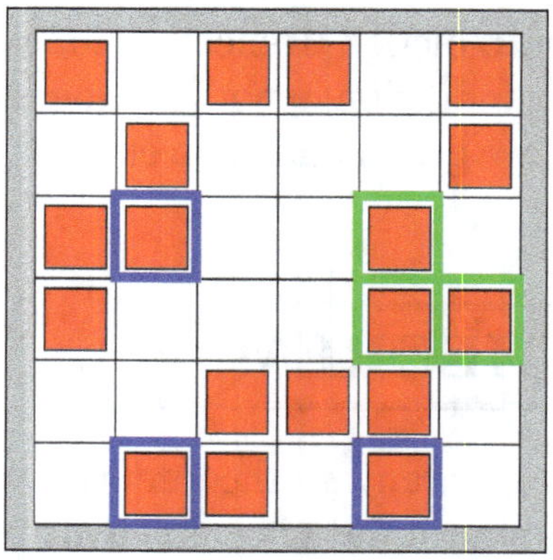

This figure shows a 6×6 grid and a subset with $\frac{1}{2}$ of the points marked with red. This selection of points contains a total of 2 corners, which are marked in green and blue, respectively.

In mathematics, the corners theorem is a result in arithmetic combinatorics proved by Miklós Ajtai and Endre Szemerédi. It states that for every $\varepsilon>0$, for large enough

N, any set of at least εN^2 points in the $N \times N$ grid $\{1,\dots,N\}\times\{1,\dots,N\}$ contains a corner, i.e., a triple of points of the form $\{(x,y),(x+h,y),(x,y+h)\}$. Later, Solymosi gave a simpler proof, based on the triangle removal lemma. The corners theorem implies Roth's theorem.

A corner is a subset of $\mathbb{Z}^2$ of the form $\{(x,y),(x+h,y),(x,y+h)\}$, where $x,y,h \in \mathbb{Z}$ and $h > 0$.

Formal Statement of Corners Theorem

If A is a subset of the $N \times N$ grid $\{1,\dots,N\}\times\{1,\dots,N\}$ that contains no corner, then the size of A is $o(N^2)$. In other words, for any ε, there is a N_0 such that for any $N \geq N_0$, any corner-free subset A of $\{1,\dots,N\}\times\{1,\dots,N\}$ is smaller than εN^2.

Proof: We would first like to replace the condition $h > 0$ with $h \neq 0$. To achieve this, we consider the set $A + A \subset \{1,\dots,2N\}\times\{1,\dots,2N\}$. By the pigeonhole principle, there exists a point $c \in A + A$ such that it can be represented as $c = a + b$ for at least $\frac{|A|^2}{(2N)^2}$ pairs $a, b \in A$. We choose this point c and construct a new set $A' := A \cap (c - A)$. Observe that $|A'| \geq \frac{|A|^2}{(2N)^2}$, as the size of A' is the number of ways of writing $c = a + b$. Further observe that it suffices to show that $|A'| = o(N^2)$. A' is a subset of A, so it has no corner, i.e., no subset of the form $\{(x,y),(x+h,y),(x,y+h)\}$ for $h > 0$. But A' is also a subset of $c - A$, so it also has no anticorner, i.e., no subset of the form $\{(x,y),(x+h,y),(x,y+h)\}$ with $h > 0$. Hence, $'$ has no subset of the form $\{(x,y),(x+h,y),(x,y+h)\}$ for $h \neq 0$, which is the condition we sought.

To show $|A'| = o(N^2)$, we construct an auxiliary tripartite graph *G*. The first part has vertex set $U = \{u_1,\dots,u_N\}$, where the vertices correspond to the N vertical lines $x = i$. The second part has vertex set $V = \{v_1,\dots,v_N\}$, where the vertices correspond to the N vertical lines $y = j$. The third part has vertex set $W = \{w_1,\dots,w_{2N}\}$, where the vertices correspond to the 2N slanted lines $y = -x + k$ with slope -1. We draw an edge between two vertices if the corresponding lines intersect at a point in A'.

Let us now think about the triangles in the auxiliary graph G. For each point $x \in A'$, the vertices of G corresponding to the horizontal, vertical, and slanted lines passing through x form a triangle in G. A case check reveals that if G contained any other triangle, then there would be a corner or anticorner, so G does not contain any other triangle. With this characterization of all the triangles in G, observe that each edge of G (corresponding to an intersection of lines at some point $x \in A'$) is contained in exactly one triangle (namely the triangle with vertices corresponding to the three lines passing through $x \in A'$). It is a well-known corollary of the triangle removal lemma that a graph on n vertices in which each edge is in a unique triangle has $o(n^2)$ edges. Hence, G has $o(N^2)$ edges. But note that we can count the edges of G exactly by just counting all the intersections at points in A' – there are $3|A'|$ such intersections. Hence, $3|A|'=|E(G)|=o(N^2)$, from which $|A'|=o(N^2)$. This completes the proof.

A Proof of Roth's Theorem from the Corners Theorem

Roth's theorem is the special case of Szemerédi's theorem for arithmetic progressions of length 3. Roth's theorem: If $A \subseteq \{1,2,\ldots,N\}$ contains no 3-term arithmetic progression, then $|A|=o(N)$.

Proof: We have $A \subseteq \{1,2,\ldots,N\}$ that does not contain any 3-term arithmetic progression. Define the following set:

$$B=\{(x,y)\in\{1,2,\ldots,2N\}\times\{1,2,\ldots,2N\} \mid x-y\in A\}.$$

For each $a\in A$, there are at least N pairs $(x,y)\in\{1,2,\ldots,2N\}\times\{1,2,\ldots,2N\}$ such that $x-y=a$. For different $a_1,a_2\in A$, these corresponding pairs are clearly different. Hence, $|B|\geq N|A|$.

Say for a contradiction that B contains a corner $\{(x,y),(x+h,y),(x,y+h)\}$. Then A contains the elements $x-(y+h), x-y, (x+h)-y$, which form a 3-term arithmetic progression – a contradiction. Hence, B is corner-free, so by the corners theorem, $|B|=o(N^2)$. Putting everything together, we have $A\leq|B|/N=o(N^2)/N=o(N)$, which is what we set out to prove.

BERTRAND'S BALLOT THEOREM

In combinatorics, Bertrand's ballot problem is the question: "In an election where candidate A receives p votes and candidate B receives q votes with p > q, what is the probability that A will be strictly ahead of B throughout the count?" The answer is:

$$\frac{p-q}{p+q}.$$

In Bertrand's original paper, he sketches a proof based on a general formula for the number of favourable sequences using a recursion relation. He remarks that it seems probable that such a simple result could be proved by a more direct method. Such a proof was given by Désiré André, based on the observation that the unfavourable sequences can be divided into two equally probable cases, one of which (the case where B receives the first vote) is easily computed; he proves the equality by an explicit bijection. A variation of his method is popularly known as André's reflection method, although André did not use any reflections.

Example: Suppose there are 5 voters, of whom 3 vote for candidate A and 2 vote for candidate B (so p = 3 and q = 2).

There are ten possibilities for the order of the votes cast:

- AAABB
- AABAB
- ABAAB
- BAAAB
- AABBA
- ABABA
- BAABA
- ABBAA
- BABAA
- BBAAA

For the order AABAB, the tally of the votes as the election progresses is:

Candidate	A	A	B	A	B
A	1	2	2	3	3
B	0	0	1	1	2

For each column the tally for A is always larger than the tally for B so the A is always strictly ahead of B. For the order AABBA the tally of the votes as the election progresses is:

Candidate	A	A	B	B	A
A	1	2	2	2	3
B	0	0	1	2	2

For this order, B is tied with A after the fourth vote, so A is not always strictly ahead of B. Of the 10 possible orders, A is always ahead of B only for AAABB and AABAB. So the probability that A will always be strictly ahead is,

$$\frac{2}{10} = \frac{1}{5},$$

and this is indeed equal to $\frac{3-2}{3+2}$ as the theorem predicts.

Equivalent Problems

Rather than computing the probability that a random vote counting order has the desired property, one can instead compute the number of favourable counting orders, then divide by the total number of ways in which the votes could have been counted. (This is the method used by Bertrand). The total number of ways is the binomial coefficient $\binom{p+q}{p}$; Bertrand's proof shows that the number of favourable orders in which to count the votes is $\binom{p+q-1}{p-1}-\binom{p+q-1}{p}$ (though he does not give this number explicitly). And indeed after division this gives $\frac{p}{p+q}-\frac{q}{p+q}=\frac{p-q}{p+q}$.

Another equivalent problem is to calculate the number of random walks on the integers that consist of n steps of unit length, beginning at the origin and ending at the point m, that never become negative. Assuming n and m have the same parity and $n \geq m \geq 0$, this number is:

$$\binom{n}{\frac{n+m}{2}}-\binom{n}{\frac{n+m}{2}+1}=\frac{m+1}{\frac{n+m}{2}+1}\binom{n}{\frac{n+m}{2}}.$$

When $m = 0$ and n is even, this gives the Catalan number $\frac{1}{\frac{n}{2}+1}\binom{n}{\frac{n}{2}}$.

Proof by Reflection

For A to be strictly ahead of B throughout the counting of the votes, there can be no ties. Separate the counting sequences. Any sequence that begins with a vote for B must reach a tie at some point, because A eventually wins. For any sequence that begins with A and reaches a tie, reflect the votes up to the point of the first tie (so any A becomes a B, and vice versa) to obtain a sequence that begins with B. Hence every sequence that begins with A and reaches a tie is in one-to-one correspondence with a sequence that begins with B, and the probability that a sequence begins with B is $q/(p+q)$, so the probability that A always leads the vote is,

$=1-$ the probability of sequences that tie at some point.

$=1-$ the probability of sequences that tie at some point and begin with A or B.

$$=1-2\frac{q}{p+q}=\frac{p-q}{p+q}.$$

Proof by Induction

Another method of proof is by mathematical induction:

- We loosen the condition $p > q$ to $p \geq q$. Clearly, the theorem is correct when $p = q$, since in this case the first candidate will not be strictly ahead after all the votes have been counted (so the probability is 0).

Clearly the theorem is true if $p > 0$ and $q = 0$ when the probability is 1, given that the first candidate receives all the votes; it is also true when $p=q>0$.

- Assume it is true both when $p = a - 1$ and $q = b$, and when $p = a$ and $q = b - 1$, with $a > b > 0$. (We don›t need to consider the case $a = b$ here, since we have already disposed of it before). Then considering the case with $p = a$ and $q = b$, the last vote counted is either for the first candidate with probability $a/(a + b)$, or for the second with probability $b/(a + b)$. So the probability of the first being ahead throughout the count to the penultimate vote counted (and also after the final vote) is:

$$\frac{a}{(a+b)}\frac{(a-1)-b}{(a+b-1)}+\frac{b}{(a+b)}\frac{a-(b-1)}{(a+b-1)}=\frac{a-b}{a+b}.$$

- And so it is true for all p and q with $p > q > 0$.

Proof by Permutation

A simple proof is based on the beautiful Cycle Lemma of Dvoretzky and Motzkin. Call a ballot sequence dominating if A is strictly ahead of B throughout the counting of the votes. The Cycle Lemma asserts that any sequence of p A's and q B's, where $p > q$, has precisely $p-q$ dominating cyclic permutations. To see this, just arrange the given sequence of $p+q$ A's and B's in a circle and repeatedly remove adjacent pairs AB until only $p-q$ A's remain. Each of these A's was the start of a dominating cyclic permutation before anything was removed. So $p-q$ out of the $p+q$ cyclic permutations of any arrangement of p A votes and q B votes are dominating.

Bertrand's and André's Proofs

Bertrand expressed the solution as,

$$\frac{2m-\mu}{\mu}$$

where $\mu = p+q$ is the total number of voters and $m = p$ is the number of voters for the first candidate. He states that the result follows from the formula:

$$P_{m+1,\mu+1} = P_{m,\mu} + P_{m+1,\mu},$$

where $P_{m,\mu}$ is the number of favourable sequences, but "it seems probable that such a simple result could be shown in a more direct way". Indeed, a more direct proof was soon produced by Désiré André. His approach is often mistakenly labelled "the reflection principle" by modern authors but in fact uses a permutation. He shows that the "unfavourable" sequences (those that reach an intermediate tie) consist of an equal number of sequences that begin with A as those that begin with B. Every sequence that begins with B is unfavourable, and there are $\binom{p+q-1}{q-1}$ such sequences with a B followed by an arbitrary sequence of (q-1) B's and p A's. Each unfavourable sequence that begins with A can be transformed to an arbitrary sequence of (q-1) B's and p A's by finding the first B that violates the rule (by causing the vote counts to tie) and deleting it, and interchanging the order of the remaining parts. To reverse the process, take any sequence of (q-1) B's and p A's and search from the end to find where the number of A's first exceeds the number of B's, and then interchange the order of the parts and place a B in between. For example, the unfavourable sequence AABBABAA corresponds uniquely to the arbitrary sequence ABAAAAB. From this, it follows that the number of favourable sequences of p A's and q B's is,

$$\binom{p+q}{q}-2\binom{p+q-1}{q-1}=\binom{p+q}{q}\frac{p-q}{p+q}$$

and thus the required probability is,

$$\frac{p-q}{p+q}$$

as expected.

Variant: Ties Allowed

The original problem is to find the probability that the first candidate is always strictly ahead in the vote count. One may instead consider the problem of finding the probability that the second candidate is never ahead (that is, with ties are allowed). In this case, the answer is,

$$\frac{p+1-q}{p+1}.$$

The variant problem can be solved by the reflection method in a similar way to the original problem. The number of possible vote sequences is $\binom{p+q}{q}$. Call a sequence "bad" if the second candidate is ever ahead, and if the number of bad sequences can be enumerated then the number of "good" sequences can be found by subtraction and the probability can be computed.

Represent a voting sequence as a lattice path on the Cartesian plane as follows:

- Start the path at (0, 0).
- Each time a vote for the first candidate is received move right 1 unit.
- Each time a vote for the second candidate is received move up 1 unit.

Each such path corresponds to a unique sequence of votes and will end at (p, q). A sequence is 'good' exactly when the corresponding path never goes above the diagonal line y = x; equivalently, a sequence is 'bad' exactly when the corresponding path touches the line y = x + 1.

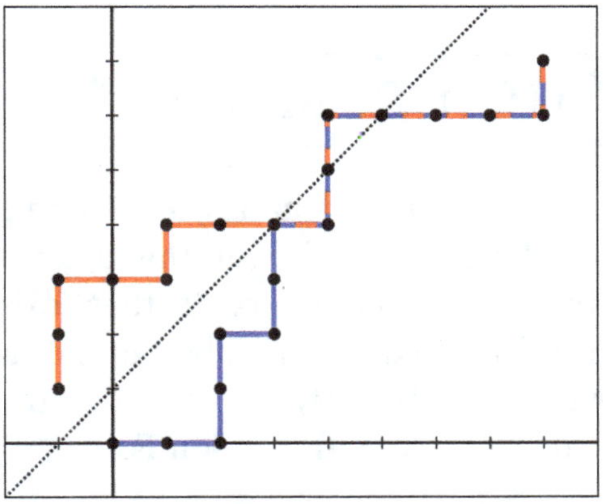

'Bad' path (blue) and its reflected path (red).

For each 'bad' path P, define a new path P′ by reflecting the part of P up to the first point it touches the line across it. P′ is a path from (−1, 1) to (p, q). The same operation applied again restores the original P. This produces a one-to-one correspondence between the 'bad' paths and the paths from (−1, 1) to (p, q). The number of these paths is $\binom{p+q}{q-1}$ and so that is the number of 'bad' sequences. This leaves the number of 'good' sequences as,

$$\binom{p+q}{q}-\binom{p+q}{q-1}=\binom{p+q}{q}\frac{p+1-q}{p+1}.$$

Since there are $\binom{p+q}{q}$ altogether, the probability of a sequence being good is $\frac{p+1-q}{p+1}$.

In fact, the solutions to the original problem and the variant problem are easily related. For candidate A to be strictly ahead throughout the vote count, they must receive the first vote and for the remaining votes (ignoring the first) they must be either strictly ahead or tied throughout the count. Hence the solution to the original problem is,

$$\frac{p}{p+q}\frac{p-1+1-q}{p-1+1}=\frac{p-q}{p+q}$$

as required.

Conversely, the tie case can be derived from the non-tie case. The number of non-tie sequences with p+1 votes for A is equal to the number of tie sequences with p votes for A. The number of non-tie votes with p + 1 votes for A votes is $\frac{p+1-q}{p+1+q}\binom{p+1+q}{q}$, which by algebraic manipulation is $\frac{p+1-q}{p+1}\binom{p+q}{q}$, so the fraction of sequences with p votes for A votes is $\frac{p+1-q}{p+1}$.

FOLKMAN'S THEOREM

Folkman's theorem is a theorem in mathematics, and more particularly in arithmetic combinatorics and Ramsey theory. According to this theorem, whenever the natural numbers are partitioned into finitely many subsets, there exist arbitrarily large sets of numbers all of whose sums belong to the same subset of the partition. The theorem had been discovered and proved independently by several mathematicians, before it was named "Folkman's theorem", as a memorial to Jon Folkman, by Graham, Rothschild, and Spencer.

Let N be the set {1, 2, 3,...} of positive integers, and suppose that N is partitioned into k different subsets N_1, N_2,... N_k, where k is any positive integer. Then Folkman's theorem states that, for every positive integer m, there exists a set S_m and an index i_m such that S_m has m elements and such that every sum of a nonempty subset of S_m belongs to N_{im}.

Relation to Rado's Theorem and Schur's Theorem

Schur's theorem in Ramsey theory states that, for any finite partition of the positive integers, there exist three numbers x, y, and x + y that all belong to the same partition set. That is, it is the special case m = 2 of Folkman's theorem.

Rado's theorem in Ramsey theory concerns a similar problem statement in which the integers are partitioned into finitely many subsets; the theorem characterizes the integer matrices A with the property that the system of linear equations A x = 0 can be guaranteed to have a solution in which every coordinate of the solution vector x belongs to the same subset of the partition. A system of equations is said to be regular whenever it satisfies the conditions of Rado's theorem; Folkman's theorem is equivalent to the regularity of the system of equations,

$$x_T = \sum_{i \in T} x_{\{i\}},$$

where T ranges over each nonempty subset of the set {1, 2,..., m}.

Multiplication versus Addition

It is possible to replace addition by multiplication in Folkman's theorem: if the natural numbers are finitely partitioned, there exist arbitrarily large sets S such that all products of nonempty subsets of S belong to a single partition set. Indeed, if one restricts S to consist only of powers of two, then this result follows immediately from the additive version of Folkman's theorem. However, it is open whether there exist arbitrarily large sets such that all sums and all products of nonempty subsets belong to a single partition set. The first example of nonlinearity in Ramsey Theory which does not consist of monomials was given, independently, by Furstenberg and Sarkozy, with the family {x, x + y²}. In 2016, J. Moreira proved there exists a set of the form {x, x + y, xy} contained in an element of the partition However it is not even known whether there must necessarily exist a set of the form {x, y, x + y, xy} for which all four elements belong to the same partition set.

Canonical Folkman Theorem

Let $FS(\{x_i\}_{i=1}^{n})$ denote the set of all finite sums of elements of $\{x_i\}_{i=1}^{n}$. Let C be a (possibly infinite) coloring of the positive integers, and let n be an arbitrary positive integer. There exists $\{x_i\}_{i=1}^{n}$ such that at least one of the following 3 conditions holds.

- $FS(\{x_i\}_{i=1}^{n})$ is a monochromatic set.
- $FS(\{x_i\}_{i=1}^{n})$ is a rainbow set.
- For any $B \subseteq [1,n]$, the color of $\sum_{i \in B} x_i$ is determined solely by min(B).

LABELLED ENUMERATION THEOREM

In combinatorial mathematics, the labelled enumeration theorem is the counterpart of the Pólya enumeration theorem for the labelled case, where we have a set of labelled objects given by an exponential generating function (EGF) g(z) which are being distributed into n slots and a permutation group G which permutes the slots, thus creating equivalence classes of configurations. There is a special re-labelling operation that re-labels the objects in the slots, assigning labels from 1 to k, where k is the total number of nodes, i.e. the sum of the number of nodes of the individual objects. The EGF $f_n(z)$ of the number of different configurations under this re-labelling process is given by,

$$f_n(z) = \frac{g(z)^n}{|G|}.$$

In particular, if G is the symmetric group of order n (hence, |G| = n!), the functions f_n(z) can be further combined into a single generating function:

$$F(z,t) = \sum_{n=0}^{\infty} f_n(z)t^n = \sum_{n=0}^{\infty} \frac{g(z)^n}{n!} t^n = e^{g(z)t}$$

which is exponential w.r.t. the variable z and ordinary w.r.t. the variable t.

The Re-labelling Process

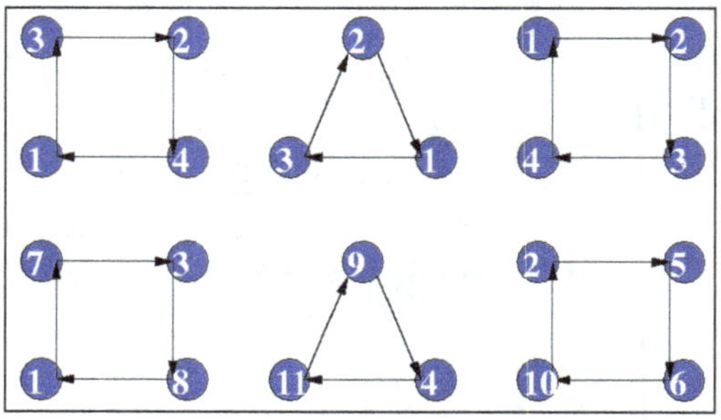

A set of cycles being re-labelled to form a permutation. (There are three slots and $G = S_3$).

We assume that an object ω of size $|\omega|$ represented by $z^{|\omega|}/|\omega|!$ contains $|\omega| = m$ labelled internal nodes, with the labels going from 1 to m. The action of G on the slots is greatly simplified compared to the unlabelled case, because the labels distinguish the objects in the slots, and the orbits under G all have the same size $|G|$. (The EGF g(z) may not include objects of size zero. This is because they are not distinguished by labels and therefore the presence of two or more of such objects creates orbits whose size is less than $|G|$) the nodes of the objects are re-labelled when they are distributed into the slots. Say an object of size r_1 goes into the first slot, an object of size r_2 into the second slot, and so on, and the total size of the configuration is k, so that,

$$r_1 + r_2 + \cdots + r_n = k.$$

The re-labelling process works as follows: choose one of,

$$\binom{k}{r_1, r_2, \ldots r_n}$$

partitions of the set of k labels into subsets of size $r_1, r_2, \ldots r_n$. Now re-label the internal nodes of each object using the labels from the respective subset, preserving the order of the labels. E.g. if the first object contains four nodes labelled from 1 to 4 and the set of labels chosen for this object is {2, 5, 6, 10}, then node 1 receives the label 2, node 2, the label 5, node 3, the label 6 and node 4, the label 10. In this way the labels on the objects induce a unique labelling using the labels from the subset of $[k]$ chosen for the object.

Proof of the Theorem

It follows from the re-labelling construction that there are:

$$\frac{1}{|G|} \sum_{r_1 + r_2 + \ldots + r_n = k} \binom{k}{r_1, r_2, \ldots r_n} r_1![z^{r_1}]g(z)\, r_2![z^{r_2}]g(z) \cdots r_n![z^{r_n}]g(z)$$

or

$$\frac{k!}{|G|}\sum_{r_1+r_2+\ldots+r_n=k}\left[z^{r_1}\right]g(z)\left[z^{r_2}\right]g(z)\cdots\left[z^{r_n}\right]g(z)=\frac{k!}{|G|}\left[z^k\right]g(z)^n$$

different configurations of total size k. The formula evaluates to an integer because $[z^k]g(z)^n$ is zero for $k < n$ (remember that g does not include objects of size zero) and when $k \geq n$ we have $n! \mid k!$ and the order $|G|$ of G divides the order of S_n, which is $n!$, by Lagrange's theorem. The conclusion is that the EGF of the labelled configurations is given by,

$$f_n(z)=\sum_{k\geq 0}\left(\frac{k!}{|G|}[z^k]g(z)^n\right)\frac{z^k}{k!}=\frac{1}{|G|}\sum_{k\geq 0}z^k[z^k]g(z)^n=\frac{g(z)^n}{|G|}.$$

This formula could also be obtained by enumerating sequences, i.e. the case when the slots are not being permuted, and by using the above argument without the $1/|G|$-factor to show that their generating function under re-labelling is given by $g(z)^n$. Finally note that every sequence belongs to an orbit of size $|G|$, hence the generating function of the orbits is given by $g(z)^n/|G|$.

SZEMERÉDI'S THEOREM

In arithmetic combinatorics, Szemerédi's theorem is a result concerning arithmetic progressions in subsets of the integers. In 1936, Erdős and Turán conjectured that every set of integers A with positive natural density contains a k-term arithmetic progression for every k. Endre Szemerédi proved the conjecture.

A subset A of the natural numbers is said to have positive upper density if,

$$\limsup_{n\to\infty}\frac{|A\cap\{1,2,3,\ldots,n\}|}{n}>0.$$

Szemerédi's theorem asserts that a subset of the natural numbers with positive upper density contains infinitely many arithmetic progressions of length k for all positive integers k.

An often-used equivalent finitary version of the theorem states that for every positive integer k and real number $\delta \in (0,1]$, there exists a positive integer,

$$N = N(k,\delta)$$

such that every subset of {1, 2,..., N} of size at least δN contains an arithmetic progression of length k.

Another formulation uses the function $r_k(N)$, the size of the largest subset of $\{1, 2, \ldots, N\}$ without an arithmetic progression of length k. Szemerédi's theorem is equivalent to the asymptotic bound,

$$r_k(N) = o(N).$$

That is, $r_k(N)$ grows less than linearly with N. The cases k = 1 and k = 2 of Szemerédi's theorem are trivial. The case k = 3, known as Roth's theorem, was established in 1953 by Klaus Roth via an adaptation of the Hardy–Littlewood circle method. Endre Szemerédi proved the case k = 4 through combinatorics. Using an approach similar to the one he used for the case k = 3, Roth gave a second proof for this in 1972.

The general case was settled in 1975, also by Szemerédi, who developed an ingenious and complicated extension of his previous combinatorial argument for k = 4 (called "a masterpiece of combinatorial reasoning" by Erdős). Several other proofs are now known, the most important being those by Hillel Furstenberg in 1977, using ergodic theory, and by Timothy Gowers in 2001, using both Fourier analysis and combinatorics. Terence Tao has called the various proofs of Szemerédi's theorem a "Rosetta stone" for connecting disparate fields of mathematics.

Quantitative Bounds

It is an open problem to determine the exact growth rate of $r_k(N)$. The best known general bounds are:

$$CN\exp\left(-n2^{(n-1)/2}\sqrt[n]{\log N}+\frac{1}{2n}\log\log N\right)\le r_k(N)\le\frac{N}{(\log\log N)^{2^{-2^{k+9}}}},$$

where $n = [\log k]$. The lower bound is due to O'Bryant building on the work of Behrend, Rankin, and Elkin. The upper bound is due to Gowers.

For small k, there are tighter bounds than the general case. When k = 3, Bourgain, Heath-Brown, Szemerédi, and Sanders provided increasingly smaller upper bounds. The current best bounds are,

$$N2^{-\sqrt{8\log N}} \le r_3(N) \le C\frac{(\log\log N)^4}{\log N}N$$

due to O'Bryant and Bloom respectively.

For k = 4, Green and Tao proved that,

$$r_4(N) \le C\frac{N}{(\log N)^c}$$

for some c > 0.

Extensions and Generalizations

A multidimensional generalization of Szemerédi's theorem was first proven by Hillel Furstenberg and Yitzhak Katznelson using ergodic theory. Timothy Gowers, Vojtěch Rödl and Jozef Skokan with Brendan Nagle, Rödl, and Mathias Schacht, and Terence Tao provided combinatorial proofs.

Alexander Leibman and Vitaly Bergelson generalized Szemerédi's to polynomial progressions: If $A \subset \mathbb{N}$ is a set with positive upper density and $p_1(n), p_2(n), \ldots, p_k(n)$ are integer-valued polynomials such that $p_i(0) = 0$, then there are infinitely many $u, n \in \mathbb{Z}$ such that $u + p_i(n) \in A$ for all $1 \leq i \leq k$. Leibman and Bergelson's result also holds in a multidimensional setting.

The finitary version of Szemerédi's theorem can be generalized to finite additive groups including vector spaces over finite fields. The finite field analog can be used as a model for understanding the theorem in the natural numbers. The problem of obtaining bounds in the k=3 case of Szemerédi's theorem in the vector space $\mathbb{F}_3^n$ is known as the cap set problem.

The Green–Tao theorem asserts the prime numbers contain arbitrary long arithmetic progressions. It is not implied by Szemerédi's theorem because the primes have density 0 in the natural numbers. As part of their proof, Ben Green and Tao introduced a "relative" Szemerédi theorem which applies to subsets of the integers (even those with 0 density) satisfying certain pseudorandomness conditions. A more general relative Szemerédi theorem has since been given by David Conlon, Jacob Fox, and Yufei Zhao. The Erdős conjecture on arithmetic progressions would imply both Szemerédi's theorem and the Green–Tao theorem.

THEOREMS IN GRAPH THEORY

Kőnig's Theorem

In the mathematical area of graph theory, Kőnig's theorem, proved by Dénes Kőnig, describes an equivalence between the maximum matching problem and the minimum vertex cover problem in bipartite graphs.

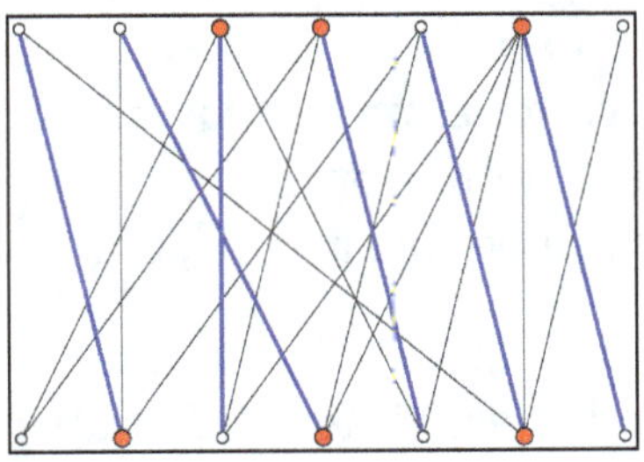

An example of a bipartite graph, with a maximum matching (blue) and minimum vertex cover (red) both of size six.

Setting

A graph is bipartite if its vertices can be partitioned into two sets such that each edge has one endpoint in each set. A vertex cover in a graph is a set of vertices that includes at least one endpoint of every edge, and a vertex cover is minimum if no other vertex cover has fewer vertices. A matching in a graph is a set of edges no two of which share an endpoint, and a matching is maximum if no other matching has more edges. Kőnig's theorem states that, in any bipartite graph, the number of edges in a maximum matching is equal to the number of vertices in a minimum vertex cover.

For graphs that are not bipartite, the maximum matching and minimum vertex cover problems are very different in complexity: maximum matchings can be found in polynomial time for any graph, while minimum vertex cover is NP-complete. The complement of a vertex cover in any graph is an independent set, so a minimum vertex cover is complementary to a maximum independent set; finding maximum independent sets is another NP-complete problem. The equivalence between matching and covering articulated in Kőnig's theorem allows minimum vertex covers and maximum independent sets to be computed in polynomial time for bipartite graphs, despite the NP-completeness of these problems for more general graph families. Kőnig's theorem is equivalent to numerous other min-max theorems in graph theory and combinatorics, such as Hall's marriage theorem and Dilworth's theorem. Since bipartite matching is a special case of maximum flow, the theorem also results from the max-flow min-cut theorem. In any bipartite graph, the number of edges in a maximum matching equals the number of vertices in a minimum vertex cover.

Example: The bipartite graph shown in the above illustration has 14 vertices; a matching with six edges is shown in blue, and a vertex cover with six vertices is shown in red. There can be no smaller vertex cover, because any vertex cover has to include at least one endpoint of each matched edge (as well as of every other edge), so this is a minimum vertex cover. Similarly, there can be no larger matching, because any matched edge has to include at least one endpoint in the vertex cover, so this is a maximum matching. Kőnig's theorem states that the equality between the sizes of the matching and the cover (in this example, both numbers are six) applies more generally to any bipartite graph.

Proof: Kőnig's theorem can be proven in a way that provides additional useful information beyond just its truth: the proof provides a way of constructing a minimum vertex cover from a maximum matching. Let G = (V,E) be a bipartite graph, and let the vertex set V be partitioned into left set L and right set R. Suppose that M is a maximum matching for G. No vertex in a vertex cover can cover more than one edge of M (because the edge half-overlap would prevent M from being a matching in the first place), so if a vertex cover with |M| vertices can be constructed, it must be a minimum cover.

To construct such a cover, let U be the set of unmatched vertices in L (possibly empty), and let Z be the set of vertices that are either in U or are connected to U by alternating paths (paths that alternate between edges that are in the matching and edges that are not in the matching).

Let,

$$K = (L \setminus Z) \cup (R \cap Z).$$

Every edge e in E either belongs to an alternating path (and has a right endpoint in K), or it has a left endpoint in K. For, if e is matched but not in an alternating path, then its left endpoint cannot be in an alternating path (because two matched edges cannot share a vertex) and thus belongs to L\Z. Alternatively, if e is unmatched but not in an alternating path, then its left endpoint cannot be in an alternating path, for such a path could be extended by adding e to it. Thus, K forms a vertex cover.

Additionally, every vertex in K is an endpoint of a matched edge. For, every vertex in L\Z is matched because Z is a superset of U, the set of unmatched left vertices. And every vertex in $R \cap Z$ must also be matched, for if there existed an alternating path to an unmatched vertex then changing the matching by removing the matched edges from this path and adding the unmatched edges in their place would increase the size of the matching. However, no matched edge can have both of its endpoints in K. Thus, K is a vertex cover of cardinality equal to M, and must be a minimum vertex cover.

Algorithm

The construction described in the proof above provides an algorithm for producing a minimum vertex cover given a maximum matching. Thus, the Hopcroft–Karp algorithm for finding maximum matchings in bipartite graphs may also be used to solve the vertex cover problem efficiently in these graphs.

Despite the equivalence of the two problems from the point of view of exact solutions, they are not equivalent for approximation algorithms. Bipartite maximum matchings can be approximated arbitrarily accurately in constant time by distributed algorithms; in contrast, approximating the minimum vertex cover of a bipartite graph requires at least logarithmic time.

Connections with Perfect Graphs

A graph is said to be perfect if, in every induced subgraph, the chromatic number equals the size of the largest clique. Any bipartite graph is perfect, because each of its subgraphs is either bipartite or independent; in a bipartite graph that is not independent the chromatic number and the size of the largest clique are both two while in an independent set the chromatic number and clique number are both one.

A graph is perfect if and only if its complement is perfect, and Kőnig's theorem can be seen as equivalent to the statement that the complement of a bipartite graph is perfect. For, each color class in a coloring of the complement of a bipartite graph is of size at most 2 and the classes of size 2 form a matching, a clique in the complement of a graph G is an independent set in G, and an independent set in a bipartite graph G is a complement of a vertex cover in G. Thus, any matching M in a bipartite graph G with n vertices corresponds to a coloring of the complement of G with n-|M| colors, which by the perfection of complements of bipartite graphs corresponds to an independent set in G with n-|M| vertices, which corresponds to a vertex cover of G with M vertices. Conversely, Kőnig's theorem proves the perfection of the complements of bipartite graphs, a result proven in a more explicit form.

One can also connect Kőnig's Line Coloring Theorem to a different class of perfect graphs, the line graphs of bipartite graphs. If G is a graph, the line graph L(G) has a vertex for each edge of G, and an edge for each pair of adjacent edges in G. Thus, the chromatic number of L(G) equals the chromatic index of G. If G is bipartite, the cliques in L(G) are exactly the sets of edges in G sharing a common endpoint. Now Kőnig's Line Coloring Theorem, stating that the chromatic index equals the maximum vertex degree in any bipartite graph, can be interpreted as stating that the line graph of a bipartite graph is perfect.

Since line graphs of bipartite graphs are perfect, the complements of line graphs of bipartite graphs are also perfect. A clique in the complement of the line graph of G is just a matching in G. And a coloring in the complement of the line graph of G, when G is bipartite, is a partition of the edges of G into subsets of edges sharing a common endpoint; the endpoints shared by each of these subsets form a vertex cover for G. Therefore, Kőnig's theorem itself can also be interpreted as stating that the complements of line graphs of bipartite graphs are perfect.

2-factor Theorem

In the mathematical discipline of graph theory, 2-factor theorem discovered by Julius Petersen, is one of the earliest works in graph theory and can be stated as follows:

> "2-factor theorem: Let G be a regular graph whose degree is an even number, 2k. Then the edges of G can be partitioned into k edge-disjoint 2-factors".

Here, a 2-factor is a subgraph of G in which all vertices have degree two; that is, it is a collection of cycles that together touch each vertex exactly once.

Proof: In order to prove this generalized form of the theorem, Petersen first proved that a 4-regular graph can be factorized into two 2-factors by taking alternate edges in a Eulerian trail. He noted that the same technique used for the 4-regular graph yields a factorization of a 2k-regular graph into two k-factors.

To prove this theorem, it is sufficient to consider connected graphs. A connected graph with even degree has an Eulerian trail. Traversing this Eulerian trail generates an orientation D of G such that every point has indegree and outdegree = k. Next, replace every vertex v ∈ V(D) by two vertices v' and v", and replace every directed edge uv of the oriented graph by an undirected edge from u' to v". Since D has in- and outdegree equal to k the resulting bipartite graph G' is k-regular. The edges of G' can be partitioned into k perfect matchings by a theorem of Kőnig. Now merging v' with v" for every v recover the graph G, and maps the k perfect matchings of G' onto k 2-factors of G which partition its edges.

Kirchhoff's Theorem

In the mathematical field of graph theory, Kirchhoff's theorem or Kirchhoff's matrix tree theorem named after Gustav Kirchhoff is a theorem about the number of spanning trees in a graph, showing that this number can be computed in polynomial time as the determinant of the Laplacian matrix of the graph. It is a generalization of Cayley's formula which provides the number of spanning trees in a complete graph.

Kirchhoff's theorem relies on the notion of the Laplacian matrix of a graph that is equal to the difference between the graph's degree matrix (a diagonal matrix with vertex degrees on the diagonals) and its adjacency matrix (a (0,1)-matrix with 1's at places corresponding to entries where the vertices are adjacent and 0's otherwise).

For a given connected graph G with n labeled vertices, let $\lambda_1, \lambda_2, \ldots, \lambda_{n-1}$ be the non-zero eigenvalues of its Laplacian matrix. Then the number of spanning trees of G is,

$$t(G) = \frac{1}{n}\lambda_1\lambda_2\cdots\lambda_{n-1}.$$

Equivalently the number of spanning trees is equal to any cofactor of the Laplacian matrix of G.

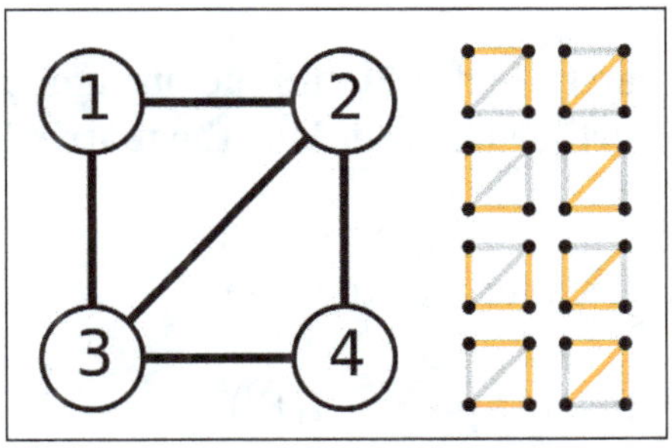

The Matrix-Tree Theorem can be used to compute the number of labeled spanning trees of this graph.

First, construct the Laplacian matrix Q for the example diamond graph G:

$$Q = \begin{bmatrix} 2 & -1 & -1 & 0 \\ -1 & 3 & -1 & -1 \\ -1 & -1 & 3 & -1 \\ 0 & -1 & -1 & 2 \end{bmatrix}.$$

Next, construct a matrix Q^* by deleting any row and any column from Q. For example, deleting row 1 and column 1 yields,

$$Q^* = \begin{bmatrix} 3 & -1 & -1 \\ -1 & 3 & -1 \\ -1 & -1 & 2 \end{bmatrix}.$$

Finally, take the determinant of Q^* to obtain t(G), which is 8 for the diamond graph. (t(G) is the (1,1)-cofactor of Q in this example).

Proof outline: First the Laplacian matrix has the property that the sum of its entries across any row and any column is 0. Thus we can transform any minor into any other minor by adding rows and columns, switching them, and multiplying a row or a column by −1. Thus the cofactors are the same up to sign, and it can be verified that, in fact, they have the same sign.

We proceed to show that the determinant of the minor M_{11} counts the number of spanning trees. Let n be the number of vertices of the graph, and m the number of its edges. The incidence matrix E is an n-by-m matrix, which may be defined as follows: suppose that (i, j) is the kth edge of the graph, and that $i < j$. Then $E_{ik} = 1$, $E_{jk} = -1$, and all other entries in column k are 0. For the preceding example (with n = 4 and m = 5):

$$E = \begin{bmatrix} 1 & 1 & 0 & 0 & 0 \\ -1 & 0 & 1 & 1 & 0 \\ 0 & -1 & -1 & 0 & 1 \\ 0 & 0 & 0 & -1 & -1 \end{bmatrix}.$$

Recall that the Laplacian L can be factored into the product of the incidence matrix and its transpose, i.e., $L = EE^T$. Furthermore, let F be the matrix E with its first row deleted, so that $FF^T = M_{11}$.

Now the Cauchy-Binet formula allows us to write,

$$\det(M_{11}) = \sum_S \det(F_S)\det(F_S^T) = \sum_S \det(F_S)^2$$

where S ranges across subsets of [m] of size n − 1, and F_S denotes the (n − 1)-by-(n − 1) matrix whose columns are those of F with index in S. Then every S specifies n − 1 edges of the original graph, and it can be shown that those edges induce a spanning tree iff the determinant of F_S is +1 or −1, and that they do not induce a spanning tree iff the determinant is 0. This completes the proof.

Particular Cases and Generalizations

Cayley's Formula

Cayley's formula follows from Kirchhoff's theorem as a special case, since every vector with 1 in one place, −1 in another place, and 0 elsewhere is an eigenvector of the Laplacian matrix of the complete graph, with the corresponding eigenvalue being n. These vectors together span a space of dimension n − 1, so there are no other non-zero eigenvalues.

Alternatively, note that as Cayley's formula counts the number of distinct labeled trees of a complete graph K_n we need to compute any cofactor of the Laplacian matrix of K_n. The Laplacian matrix in this case is,

$$\begin{bmatrix} n-1 & -1 & \cdots & -1 \\ -1 & n-1 & \cdots & -1 \\ \vdots & \vdots & \ddots & \vdots \\ -1 & -1 & \cdots & n-1 \end{bmatrix}.$$

Any cofactor of the above matrix is n^{n-2}, which is Cayley's formula.

Kirchhoff's Theorem for Multigraphs

Kirchhoff's theorem holds for multigraphs as well; the matrix Q is modified as follows:

- The entry $q_{i,j}$ equals $-m$, where m is the number of edges between i and j.
- When counting the degree of a vertex, all loops are excluded.

Cayley's formula for a complete multigraph is $m^{n-1}(n^{n-1}-(n-1)n^{n-2})$ by same methods produced above, since a simple graph is a multigraph with $m = 1$.

Explicit Enumeration of Spanning Trees

Kirchhoff's theorem can be strengthened by altering the definition of the Laplacian matrix. Rather than merely counting edges emanating from each vertex or connecting a pair of vertices, label each edge with an indeterminate and let the (i, j)-th entry of the modified Laplacian matrix be the sum over the indeterminates corresponding to edges between the i-th and j-th vertices when i does not equal j, and the negative sum over all indeterminates corresponding to edges emanating from the i-th vertex when i equals j.

The determinant of the modified Laplacian matrix by deleting any row and column (similar to finding the number of spanning trees from the original Laplacian matrix), above is then a homogeneous polynomial (the Kirchhoff polynomial) in the indeterminates corresponding to the edges of the graph. After collecting terms and performing all possible cancellations, each monomial in the resulting expression represents a spanning tree consisting of the edges corresponding to the indeterminates appearing in that monomial. In this way, one can obtain explicit enumeration of all the spanning trees of the graph simply by computing the determinant.

Matroids

The spanning trees of a graph form the bases of a graphic matroid, so Kirchhoff's theorem provides a formula to count the number of bases in a graphic matroid. The same method may also be used to count the number of bases in regular matroids, a generalization of the graphic matroids.

Kirchhoff's Theorem for Directed Multigraphs

Kirchhoff's theorem can be modified to count the number of oriented spanning trees in directed multigraphs. The matrix Q is constructed as follows:

- The entry $q_{i,j}$ for distinct i and j equals −m, where m is the number of edges from i to j.
- The entry $q_{i,i}$ equals the indegree of i minus the number of loops at i.

The number of oriented spanning trees rooted at a vertex i is the determinant of the matrix gotten by removing the ith row and column of Q.

Wagner's Theorem

In graph theory, Wagner's theorem is a mathematical forbidden graph characterization of planar graphs, named after Klaus Wagner, stating that a finite graph is planar if and only if its minors include neither K_5 (the complete graph on five vertices) nor $K_{3,3}$ (the utility graph, a complete bipartite graph on six vertices). This was one of the earliest results in the theory of graph minors and can be seen as a forerunner of the Robertson–Seymour theorem.

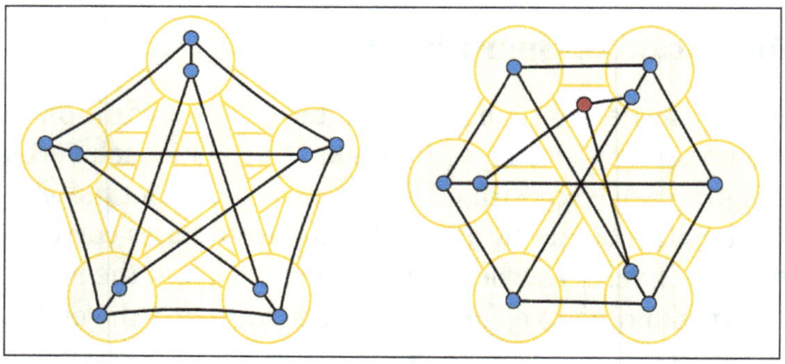

K_5 (left) and $K_{3,3}$ (right) as minors of the nonplanar Petersen graph (small colored circles and solid black edges). The minors may be formed by deleting the red vertex and contracting edges within each yellow circle.

A planar embedding of a given graph is a drawing of the graph in the Euclidean plane, with points for its vertices and curves for its edges, in such a way that the only intersections between pairs of edges are at a common endpoint of the two edges. A minor of a given graph is another graph formed by deleting vertices, deleting edges, and contracting edges. When an edge is contracted, its two endpoints are merged to form a single vertex. In some versions of graph minor theory the graph resulting from a contraction is simplified by removing self-loops and multiple adjacencies, while in other version multigraphs are allowed, but this variation makes no difference to Wagner's theorem. Wagner's theorem states that every graph has either a planar embedding, or a minor of one of two types, the complete graph K_5 or the complete bipartite graph $K_{3,3}$. (It is also possible for a single graph to have both types of minor).

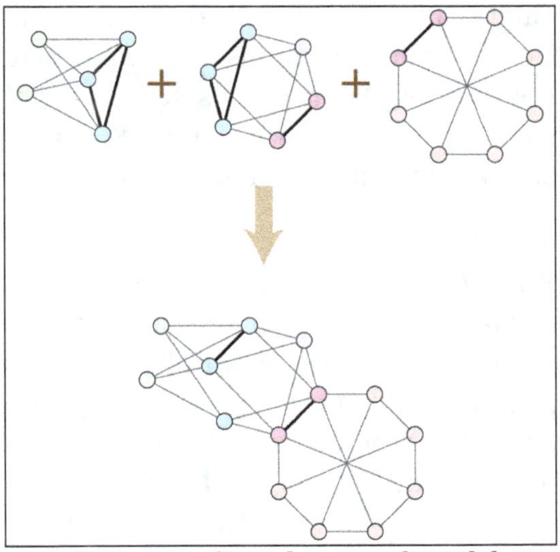

A clique-sum of two planar graphs and the Wagner graph, forming a K_5-free graph.

If a given graph is planar, so are all its minors: vertex and edge deletion obviously preserve planarity, and edge contraction can also be done in a planarity-preserving way, by leaving one of the two endpoints of the contracted edge in place and routing all of the edges that were incident to the other endpoint along the path of the contracted edge. A minor-minimal non-planar graph is a graph that is not planar, but in which all proper minors (minors formed by at least one deletion or contraction) are planar. Another way of stating Wagner's theorem is that there are only two minor-minimal non-planar graphs, K_5 and $K_{3,3}$.

Another result also sometimes known as Wagner's theorem states that a four-connected graph is planar if and only if it has no K_5 minor. That is, by assuming a higher level of connectivity, the graph $K_{3,3}$ can be made unnecessary in the characterization, leaving only a single forbidden minor, K_5. Correspondingly, the Kelmans–Seymour conjecture states that a 5-connected graph is planar if and only if it does not have K_5 as a topological minor.

Wagner published both theorems in 1937, subsequent to the 1930 publication of Kuratowski's theorem, according to which a graph is planar if and only if it does not contain as a subgraph a subdivision of one of the same two forbidden graphs K_5 and $K_{3,3}$. In a sense, Kuratowski's theorem is weaker than Wagner's theorem: a subdivision can be converted into a minor of the same type by contracting all but one edge in each path formed by the subdivision process, but converting a minor into a subdivision of the same type is not always possible. However, in the case of the two graphs K_5 and $K_{3,3}$, it is straightforward to prove that a graph that has at least one of these two graphs as a minor also has at least one of them as a subdivision, so the two theorems are equivalent.

Implications

One consequence of the stronger version of Wagner's theorem for four-connected graphs is to characterize the graphs that do not have a K_5 minor. The theorem can be rephrased as stating that every such graph is either planar or it can be decomposed into simpler pieces. Using this idea, the K_5-minor-free graphs may be characterized as the graphs that can be formed as combinations of planar graphs and the eight-vertex Wagner graph, glued together by clique-sum operations. For instance, $K_{3,3}$ can be formed in this way as a clique-sum of three planar graphs, each of which is a copy of the tetrahedral graph K_4.

Wagner's theorem is an important precursor to the theory of graph minors, which culminated in the proofs of two deep and far-reaching results: the graph structure theorem (a generalization of Wagner's clique-sum decomposition of K_5-minor-free graphs) and the Robertson–Seymour theorem (a generalization of the forbidden minor characterization of planar graphs, stating that every graph family closed under the operation of taking minors has a characterization by a finite number of forbidden minors). Analogues of Wagner's theorem can also be extended to the theory of matroids: in particular, the same two graphs K_5 and $K_{3,3}$ (along with three other forbidden configurations) appear in a characterization of the graphic matroids by forbidden matroid minors.

PÓLYA ENUMERATION THEOREM

The Pólya enumeration theorem, also known as the Redfield–Pólya theorem and Pólya counting, is a theorem in combinatorics that both follows from and ultimately generalizes Burnside's lemma on the number of orbits of a group action on a set. The theorem was first published by John Howard Redfield in 1927. In 1937 it was independently rediscovered by George Pólya, who then greatly popularized the result by applying it to many counting problems, in particular to the enumeration of chemical compounds. The Pólya enumeration theorem can also be incorporated into symbolic combinatorics and the theory of combinatorial species.

Simplified and Unweighted Version

Let X be a finite set and let G be a group of permutations of X (or a finite symmetry group that acts on X). The set X may represent a finite set of beads, and G may be a chosen group of permutations of the beads. For example, if X is a necklace of n beads in a circle, then rotational symmetry is relevant so G is the cyclic group C_n, while if X is a bracelet of n beads in a circle, rotations and reflections are relevant so G is the dihedral group D_n of order 2n. Suppose further that Y is a finite set of colors — the colors of the beads — so that Y^X is the set of colored arrangements of beads (more formally: Y^X is the set of functions $X \to Y$). Then the group G acts on Y^X. The Pólya enumeration theorem counts the number of orbits under G of colored arrangements of beads by the following formula:

$$|Y^X / G| = \frac{1}{|G|} \sum_{g \in G} m^{c(g)}$$

where $m = |Y|$ is the number of colors and c(g) is the number of cycles of the group element g when considered as a permutation of X.

Full and Weighted Version

In the more general and more important version of the theorem, the colors are also weighted in one or more ways, and there could be an infinite number of colors provided that the set of colors has a generating function with finite coefficients. In the univariate case, suppose that:

$$f(t) = f_0 + f_1 t + f_2 t^2 + \cdots$$

is the generating function of the set of colors, so that there are f_w colors of weight w for each integer $w \geq 0$. In the multivariate case, the weight of each color is a vector of integers and there is a generating function $f(t_1, t_2, ...)$ that tabulates the number of colors with each given vector of weights.

The enumeration theorem employs another multivariate generating function called the cycle index:

$$Z_G(t_1, t_2, \ldots, t_n) = \frac{1}{|G|} \sum_{g \in G} t_1^{c_1(g)} t_2^{c_2(g)} \cdots t_n^{c_n(g)}$$

where n is the number of elements of X and $c_k(g)$ is the number of k-cycles of the group element g as a permutation of X.

A colored arrangement is an orbit of the action of G on the set Y^X (where Y is the set of colors and Y^X denotes the set of all functions $\varphi: X \to Y$). The weight of such an arrangement is defined as the sum of the weights of $\varphi(x)$ over all x in X. The theorem states that the generating function F of the number of colored arrangements by weight is given by:

$$F(t) = Z_G(f(t), f(t^2), f(t^3), \ldots, f(t^n))$$

or in the multivariate case:

$$F(t_1, t_2, \ldots) = Z_G(f(t_1, t_2, \ldots), f(t_1^2, t_2^2, \ldots), f(t_1^3, t_2^3, \ldots), \ldots, f(t_1^n, t_2^n, \ldots)).$$

To reduce to the simplified version given earlier, if there are m colors and all have weight 0, then $f(t) = m$ and

$$|Y^X / G| = F(0) = Z_G(m, m, \ldots, m) = \frac{1}{|G|} \sum_{g \in G} m^{c(g)}.$$

In the celebrated application of counting trees and acyclic molecules, an arrangement of "colored beads" is actually an arrangement of arrangements, such as branches of a rooted tree. Thus the generating function f for the colors is derived from the generating function F for arrangements, and the Pólya enumeration theorem becomes a recursive formula.

Colored Cubes

How many ways are there to color the sides of a three-dimensional cube with m colors, up to rotation of the cube? The rotation group C of the cube acts on the six sides of the cube, which are equivalent to beads. Its cycle index is,

$$Z_C(t_1,t_2,t_3,t_4)=\frac{1}{24}\left(t_1^6+6t_1^2t_4+3t_1^2t_2^2+8t_3^2+6t_2^3\right)$$

which is obtained by analyzing the action of each of the 24 elements of C on the 6 sides of the cube.

We take all colors to have weight 0 and find that there are:

$$F(0)=Z_C(m,m,m,m)=\frac{1}{24}\left(m^6+3m^4+12m^3+8m^2\right)$$

different colorings.

Graphs on Three and Four Vertices

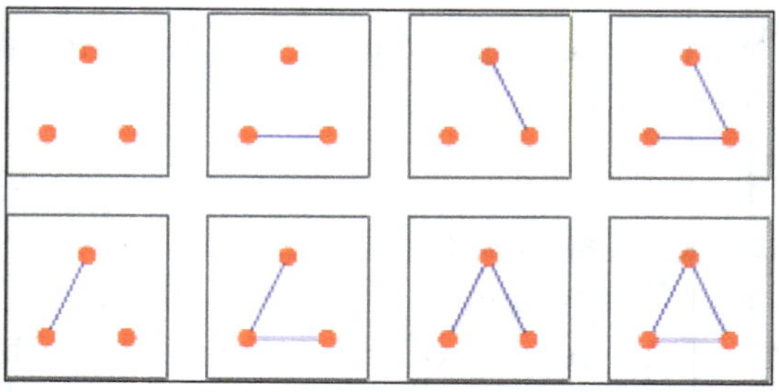

All graphs on three vertices.

A graph on m vertices can be interpreted as an arrangement of colored beads. The set X of "beads" is the set of $\binom{m}{2}$ possible edges, while the set of colors Y = {black,white} corresponds to edges that are present (black) or absent (white). The Pólya enumeration theorem can be used to calculate the number of graphs up to isomorphism with a fixed number of vertices, or the generating function of these graphs according to the number of edges they have. For the latter purpose, we can say that a black or present edge has weight 1, while an absent or white edge has weight 0. Thus $f(t)=1+t$ is the generating function for the set of colors. The relevant symmetry group is $G=S_m$, the symmetric group on m letters. This group acts on the set X of possible edges: a permutation φ turns the edge {a,b} into the edge {φ(a), φ(b)}. With these definitions, an isomorphism class of graphs with m vertices is the same as an orbit of the action of G on the set Y^X of colored arrangements; the number of edges of the graph equals the weight of the arrangement.

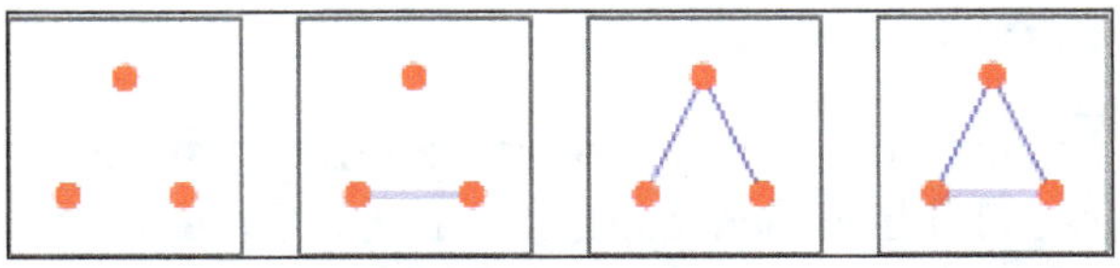

Nonisomorphic graphs on three vertices.

The eight graphs on three vertices (before identifying isomorphic graphs) are shown at the right. There are four isomorphism classes of graphs. The cycle index of the group S_3 acting on the set of three edges is,

$$Z_G(t_1,t_2,t_3)=\frac{1}{6}\left(t_1^3+3t_1t_2+2t_3\right)$$

(obtained by inspecting the cycle structure of the action of the group elements). Thus, according to the enumeration theorem, the generating function of graphs on 3 vertices up to isomorphism is,

$$F(t)=Z_G(t+1,t^2+1,t^3+1)=\frac{1}{6}\left((t+1)^3+3(t+1)(t^2+1)+2(t^3+1)\right),$$

which simplifies to,

$$F(t)=t^3+t^2+t+1.$$

Thus there is one graph each with 0 to 3 edges.

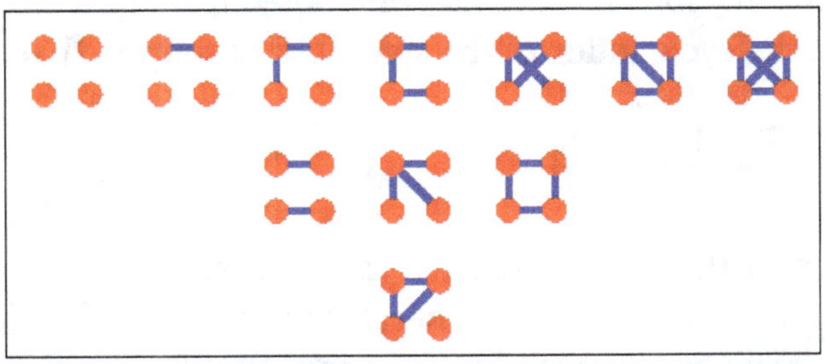

Isomorphism classes of graphs on four vertices.

The cycle index of the group S_4 acting on the set of 6 edges is,

$$Z_G(t_1,t_2,t_3,t_4)=\frac{1}{24}\left(t_1^6+9t_1^2t_2^2+8t_3^2+6t_2t_4\right)$$

Hence,

$$F(t)=Z_G(t+1,t^2+1,t^3+1,t^4+1)=\frac{(t+1)^6+9(t+1)^2(t^2+1)^2+8(t^3+1)^2+6(t^2+1)(t^4+1)}{24}$$

which simplifies to,

$$F(t)=t^6+t^5+2t^4+3t^3+2t^2+t+1.$$

Rooted Ternary Trees

The set T_3 of rooted ternary trees consists of rooted trees where every node (or non-leaf vertex) has exactly three children (leaves or subtrees). Small ternary trees are shown at right. Note that rooted ternary trees with n nodes are equivalent to rooted trees with n vertices of degree at most 3 (by ignoring the leaves). In general, two rooted trees are isomorphic when one can be obtained from the other by permuting the children of its nodes. In other words, the group that acts on the children of a node is the symmetric group S_3. We define the weight of such a ternary tree to be the number of nodes (or non-leaf vertices).

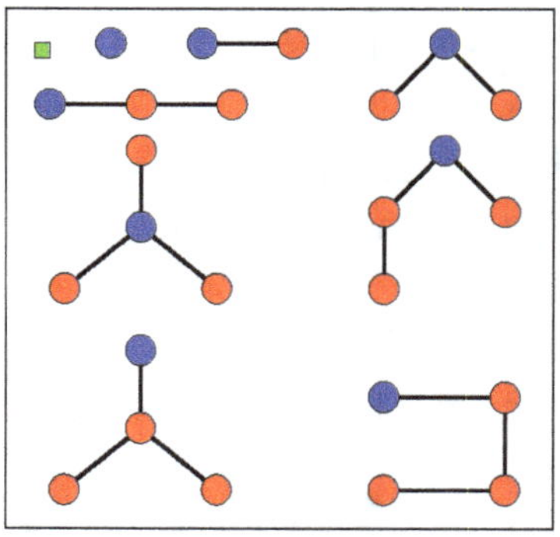

Rooted ternary trees on 0, 1, 2, 3 and 4 nodes (=non-leaf vertices). The root is shown in blue, the leaves are not shown. Every node has as many leaves as to make the number of its children equal to 3.

One can view a rooted, ternary tree as a recursive object which is either a leaf or a node with three children which are themselves rooted ternary trees. These children are equivalent to beads; the cycle index of the symmetric group S_3 that acts on them is,

$$Z_{S_3}(t_1,t_2,t_3)=\frac{t_1^3+3t_1t_2+2t_3}{6}.$$

The Polya enumeration theorem translates the recursive structure of rooted ternary trees into a functional equation for the generating function F(t) of rooted ternary trees by number of nodes. This is achieved by "coloring" the three children with rooted ternary trees, weighted by node number, so that the color generating function is given by,

$$f(t)=F(t)$$

which by the enumeration theorem gives,

$$\frac{F(t)^3+3F(t)F(t^2)+2F(t^3)}{6}$$

as the generating function for rooted ternary trees, weighted by one less than the node number (since the sum of the children weights does not take the root into account), so that,

$$F(t)=1+t\frac{F(t)^3+3F(t)F(t^2)+2F(t^3)}{6}.$$

This is equivalent to the following recurrence formula for the number t_n of rooted ternary trees with n nodes: $t_0 = 1$

and

$$t_{n+1} = \frac{1}{6}\left(\sum_{a+b+c=n} t_a t_b t_c + 3 \sum_{a+2b=n} t_a t_b + 2 \sum_{3a=n} t_a \right)$$

where a, b and c are nonnegative integers.

The first few values of t_n are 1, 1, 1, 2, 4, 8, 17, 39, 89, 211, 507, 1238, 3057, 7639, 19241.

Proof of Theorem

The simplified form of the Pólya enumeration theorem follows from Burnside's lemma, which says that the number of orbits of colorings is the average of the number of elements of Y^X fixed by the permutation g of G over all permutations g. The weighted version of the theorem has essentially the same proof, but with a refined form of Burnside's lemma for weighted enumeration. It is equivalent to apply Burnside's lemma separately to orbits of different weight.

For clearer notation, let $x_1, x_2, \ldots$ be the variables of the generating function f of Y. Given a vector of weights ω, let x^ω denote the corresponding monomial term of f. Applying Burnside's lemma to orbits of weight $\grave{u}$, the number of orbits of this weight is,

$$\frac{1}{|G|} \sum_{g \in G} |(Y^X)_{\omega,g}|$$

where $(Y^X)_{\omega,g}$ is the set of colorings of weight ω that are also fixed by g. If we then sum over all possible weights, we obtain,

$$F(x_1, x_2, \ldots) = \frac{1}{|G|} \sum_{g \in G, \omega} x^\omega |(Y^X)_{\omega,g}|.$$

Meanwhile a group element g with cycle structure $j_1(g), j_2(g), \ldots, j_n(g)$ will contribute the term,

$$t_1^{j_1(g)} t_2^{j_2(g)} \cdots t_n^{j_n(g)}$$

to the cycle index of G. The element g fixes an element φ of Y^X if and only if the function φ is constant on every cycle q of g. For every such cycle q, the generating function by weight of $|q|$ identical colors from the set enumerated by f is,

$$f(x_1^{|q|}, x_2^{|q|}, x_3^{|q|}, \ldots).$$

It follows that the generating function by weight of the points fixed by g is the product of the above term over all cycles of g, i.e.

$$\sum_{\omega} x^{\omega} \,|\, (Y^X)_{\omega,g} \,| = \prod_{q \text{ cycle of } g} f(x_1^{|q|}, x_2^{|q|}, x_3^{|q|}, \ldots),$$

which equals,

$$f(x_1, x_2, \ldots)^{j_1(g)} f(x_1^2, x_2^2, \ldots)^{j_2(g)} \cdots f(x_1^n, x_2^n, \ldots)^{j_n(g)}.$$

Substituting this for $\sum_{\omega} x^{\omega} \,|\, (Y^X)_{\omega,g} \,|$ in the sum over all g yields the substituted cycle index as claimed.

HALL'S MATCHING THEOREM

Hall's marriage theorem is a result in combinatorics that specifies when distinct elements can be chosen from a collection of overlapping finite sets. It is equivalent to several beautiful theorems in combinatorics, including Dilworth's theorem.

Combinatorial Formulation

Let S be a (possibly infinite) family of finite subsets of X, where the members of S are counted with multiplicity. (That is, S may contain the same set several times). A transversal for S is the image of an injective function f from S to X such that f(s) is an element of the set s for every s in the family S. In other words, f selects one representative from each set in S in such a way that no two sets from S get the same representative. An alternative term for transversal is system of distinct representatives.

The collection S satisfies the marriage condition when for each subfamily $W \subseteq S$,

$$|\mathbf{W}| \le \left| \bigcup_{\mathbf{A} \in \mathbf{W}} \mathbf{A} \right|.$$

Restated in words, the marriage condition asserts that every subfamily W of S covers at least |W| different members of X.

If the marriage condition fails then there cannot be a transversal f of S.

Suppose that the marriage condition fails, i.e., that for some subcollection W_0 of S, $|W_0| > |\bigcup_{A \in W_0} A|$. Suppose, by way of contradiction, that a transversal f(s) of S also exists.

The restriction of f to the offending subcollection W_0 would be an injective function from W_0 into $\bigcup_{A \in W_0}$. This is impossible by the pigeonhole principle since $|W_0| > |\bigcup_{A \in W_0} A|$.

Therefore no transversal can exist if the marriage condition fails.

Hall's theorem states that the converse is also true:

Hall's Marriage Theorem: A family S of finite sets has a transversal if and only if S satisfies the marriage condition.

Example: Consider $S=\{A_1,A_2,A_3\}$ with,

$$A_1 = \{1, 2, 3\}$$

$$A_2 = \{1, 4, 5\}$$

$$A_3 = \{3, 5\}.$$

A valid transversal would be (1, 4, 5). (This is not unique: (2, 1, 3) works equally well, for example).

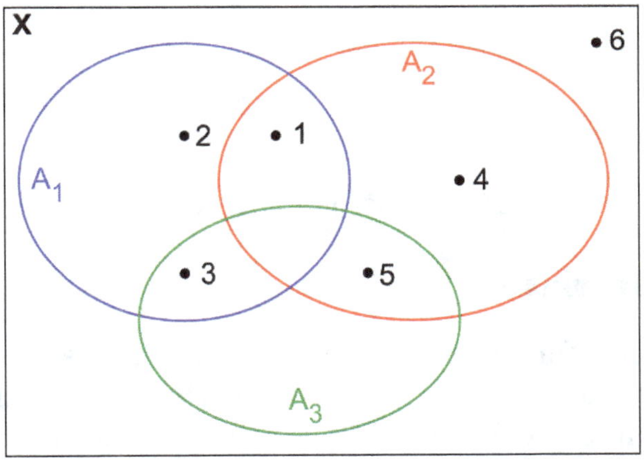

Marriage condition met.

Example: Consider $S=\{A_1,A_2,A_3,A_4\}$ with,

$$A_1 = \{2, 3, 4, 5\}$$

$$A_2 = \{4, 5\}$$

$$A_3 = \{5\}$$

$$A_4 = \{4\}.$$

No valid transversal exists; the marriage condition is violated as is shown by the subfamily $W = \{A_2, A_3, A_4\}$. Here the number of sets in the subfamily is $|W| = 3$, while the union of the three sets $A_2 \cup A_3 \cup A_4$ contains only 2 elements of X.

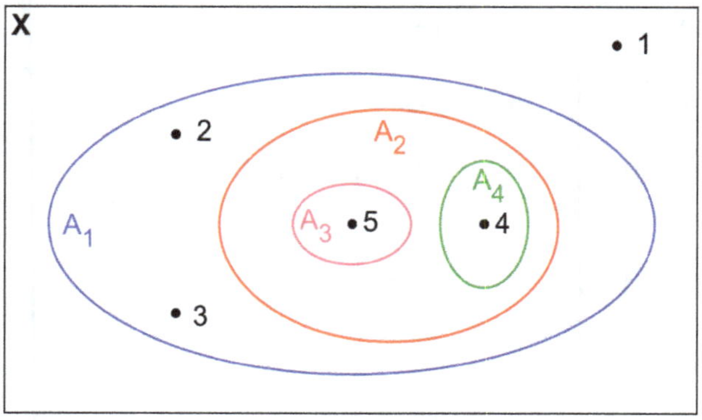

Marriage condition violated.

Example: Consider $S=\{A_1,A_2,A_3,A_4\}$ with,

$$A_1 = \{a, b, c\}$$

$$A_2 = \{b, d\}$$

$$A_3 = \{a, b, d\}$$

$$A_4 = \{b, d\}.$$

The only valid transversals are (c, b, a, d) and (c, d, a, b).

Application to Marriage

The standard example of an application of the marriage theorem is to imagine two groups; one of n men, and one of n women. For each woman, there is a subset of the men, any one of which she would happily marry; and any man would be happy to marry a woman who wants to marry him. Consider whether it is possible to pair up (in marriage) the men and women so that every person is happy.

If we let A_i be the set of men that the i-th woman would be happy to marry, then the marriage theorem states that each woman can happily marry a man if and only if the collection of sets $\{A_i\}$ meets the marriage condition. The marriage condition is that, for any subset I of the women, the number of men whom at least one of the women would be happy to marry, $|\bigcup_{i \in I} A_i|$, be at least as big as the number of women in that subset, $|I|$. It is obvious that this condition is necessary, as if it does not hold, there are not enough men to share among the I women. What is interesting is that it is also a sufficient condition.

Graph Theoretic Formulation

Let G be a finite bipartite graph with bipartite sets X and Y (i.e. G:= (X + Y, E)). An X-saturating matching is a matching which covers every vertex in X.

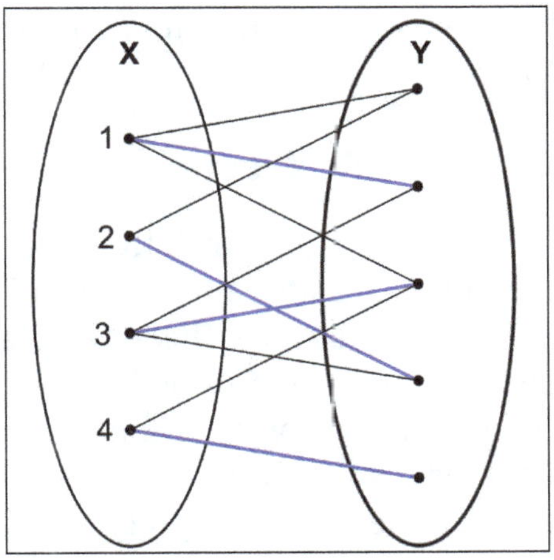

Blue edges represent a matching.

For a subset W of X, let $N_G(W)$ denote the neighborhood of W in G, i.e. the set of all vertices in Y adjacent to some element of W. The marriage theorem in this formulation states that there is an X-saturating matching if and only if for every subset W of X:

$$|W| \leq |N_G(W)|.$$

In other words: Every subset W of X has sufficiently many adjacent vertices in Y.

Proof of the Graph Theoretic Version

- Easy direction: We assume that some matching M saturates every vertex of X, and prove that Hall's condition is satisfied for all $W \subseteq X$. Let M(W) denote the set of all vertices in Y matched by M to a given W. By definition of a matching, $|M(W)| = |W|$. But $M(W) \subseteq N_G(W)$, since all elements of M(W) are neighbours of W. So, $|N_G(W)| \geq |M(W)|$ and hence, $|N_G(W)| \geq |W|$.
- Hard direction: We assume that there is no X-saturating matching and prove that Hall's condition is violated for at least one $W \subseteq X$. Let M be a maximum matching, and u a vertex not saturated by M. Consider all alternating paths (i.e., paths in G alternately using edges outside and inside M) starting from u. Let the set of all points in Y connected to u by these alternating paths be Z, and the set of all points in X connected to u by these alternating paths (including u itself) be W. No maximal alternating path can end in a vertex in Y, lest it would be an augmenting path, so that we could augment M to a strictly larger matching by toggling the status (belongs to M or not) of all the edges of the path. Thus every vertex in Z is matched by M to a vertex in $W \setminus \{u\}$. Conversely, every vertex v in $W \setminus \{u\}$ is matched by M to a vertex in Z (namely, the vertex preceding v on an alternating path ending at v). Thus, M provides a bijection of $W \setminus \{u\}$ and Z, which implies $|W| = |Z| + 1$. On the other hand, $N_G(W) \subseteq Z$: let v in Y be connected to a vertex w in W. If the edge (w,v) is in M, then v is in Z by the previous part of the proof, otherwise we can take an alternating path ending in w and

extend it with v, getting an augmenting path and showing that v is in Z. Hence, $|N_G(W)| \leq |Z| = |W| - 1 < |W|$.

Constructive Proof of the Hard Direction

Define a Hall violator as a subset W of X for which $|N_G(W)| < |W|$. If W is a Hall violator, then there is no matching that saturates all vertices of W. Therefore, there is also no matching that saturates X. Hall's marriage theorem says that a graph contains an X-saturating matching if-and-only-if it contains no Hall violators. The following algorithm proves the hard direction of the theorem: it finds either an X-saturating matching or a Hall violator. It uses, as a subroutine, an algorithm that, given a matching M and an unmatched vertex x_0, either finds an M-augmenting path or a Hall violator containing x_0.

It uses:

- Initialize M:= {}. // Empty matching.
- Assert: M is a matching in G.
- If M saturates all vertices of X, then return the X-saturating matching M.
- Let x_0 be an unmatched vertex (a vertex in $X \setminus V(M)$).
- Using the Hall violator algorithm, find either a Hall violator or an M-augmenting path.
- In the first case, return the Hall violator.
- In the second case, use the M-augmenting path to increase the size of M (by one edge), and go back to step 2.

At each iteration, M grows by one edge. Hence, this algorithm must end after at most |E| iterations. Each iteration takes at most |X| time. The total runtime complexity is similar to the Ford-Fulkerson method for finding a maximum cardinality matching.

Equivalence of the Combinatorial Formulation and the Graph-theoretic Formulation

Let $S = (A_1, A_2,\dots, A_n)$ where the A_i are finite sets which need not be distinct. Let the set $X = \{A_1, A_2,\dots, A_n\}$ (that is, the set of names of the elements of S) and the set Y be the union of all the elements in all the A_i.

We form a finite bipartite graph G:= (X + Y, E), with bipartite sets X and Y by joining any element in Y to each A_i which it is a member of. A transversal of S is an X-saturating matching (a matching which covers every vertex in X) of the bipartite graph G. Thus a problem in the combinatorial formulation can be easily translated to a problem in the graph-theoretic formulation.

Applications

The theorem has many other interesting "non-marital" applications. For example, take a standard deck of cards, and deal them out into 13 piles of 4 cards each. Then, using the marriage theorem, we can show that it is always possible to select exactly 1 card from each pile, such that the 13 selected cards contain exactly one card of each rank. More abstractly, let G be a group, and H be a finite subgroup of G. Then the marriage theorem can be used to show that there is a set T such that T is a transversal for both the set of left cosets and right cosets of H in G. The marriage theorem is used in the usual proofs of the fact that an $(r \times n)$ Latin rectangle can always be extended to an $((r+1) \times n)$ Latin rectangle when $r < n$, and so, ultimately to a Latin square.

Logical Equivalences

This theorem is part of a collection of remarkably powerful theorems in combinatorics, all of which are related to each other in an informal sense in that it is more straightforward to prove one of these theorems from another of them than from first principles. These include:

- The König–Egerváry theorem,
- König's theorem,
- Menger's theorem,
- The max-flow min-cut theorem (Ford–Fulkerson algorithm),
- The Birkhoff–Von Neumann theorem,
- Dilworth's theorem.

In particular, there are simple proofs of the implications Dilworth's theorem ⇔ Hall's theorem ⇔ König–Egerváry theorem ⇔ König's theorem.

Infinite Families

Marshall Hall Jr. Variant

By examining Philip Hall's original proof carefully, Marshall Hall, Jr. (no relation to Philip Hall) was able to tweak the result in a way that permitted the proof to work for infinite S. This variant refines the marriage theorem and provides a lower bound on the number of transversals that a given S may have. This variant is:

Suppose that $(A_1, A_2, \ldots, A_n)$, where the A_i are finite sets that need not be distinct, is a family of sets satisfying the marriage condition, and suppose that $|A_i| \geq r$ for $i = 1, \ldots, n$. Then the number of different transversals for the family is at least $r!$ if $r \leq n$ and $r(r - 1) \ldots (r - n + 1)$ if $r > n$.

Recall that a transversal for a family S is an ordered sequence, so two different transversals could have exactly the same elements. For instance, the family $A_1 = \{1,2,3\}$, $A_2 = \{1, 2, 5\}$ has both (1, 2) and (2, 1) as distinct transversals.

Marriage Condition does not Extend

The following example, due to Marshall Hall, Jr., shows that the marriage condition will not guarantee the existence of a transversal in an infinite family in which infinite sets are allowed.

Let S be the family, $A_0 = \{1, 2, 3,...\}$, $A_1 = \{1\}$, $A_2 = \{2\}$,..., $A_i = \{i\}$,...

The marriage condition holds for this infinite family, but no transversal can be constructed.

The more general problem of selecting a (not necessarily distinct) element from each of a collection of non-empty sets (without restriction as to the number of sets or the size of the sets) is permitted in general only if the axiom of choice is accepted.

The marriage theorem does extend to the infinite case if stated properly. Given a bipartite graph with sides A and B, we say that a subset C of B is smaller than or equal in size to a subset D of A in the graph if there exists an injection in the graph (namely, using only edges of the graph) from C to D, and that it is strictly smaller in the graph if in addition there is no injection in the graph in the other direction. Note that omitting in the graph yields the ordinary notion of comparing cardinalities. The infinite marriage theorem states that there exists an injection from A to B in the graph, if and only if there is no subset C of A such that N(C) is strictly smaller than C in the graph.

References

- Chartrand, Gary; Lesniak, Linda; Zhang, Ping (2010), Graphs & Digraphs (5th ed.), CRC Press, p. 307, ISBN 9781439826270
- Dilworths-theorem: geeksforgeeks.org, Retrieved 16 January, 2020
- torer, J. A. (2001), An Introduction to Data Structures and Algorithms, Progress in Computer Science and Applied Logic Series, Springer, ISBN 9780817642532
- Lovász, László (2006), "Graph minor theory", Bulletin of the American Mathematical Society, 43 (1): 75–86, doi:10.1090/S0273-0979-05-01088-8, MR 2188176

Enumerative Combinatorics

CHAPTER 4

The area of combinatorics that is concerned with the number of ways in which certain patterns can be created is called enumerative combinatorics. Generating function, alternating sign matrix, exponential formula, lattice path, etc. are a few of its concepts. All the aspects related to enumerative combinatorics have been carefully written to provide an easy understanding of the subject.

Enumerative combinatorics deals with finite sets and their cardinalities. In other words, a typical problem of enumerative combinatorics is to find the number of ways a certain pattern can be formed. The basic problem of enumerative combinatorics is that of counting the number of elements of a finite set. Usually we are given an infinite dass of finite sets S_i where i ranges over some index set I (such as the nonnegative integers $\mathbb{N}$), and we wish to count the number f(i) of elements of each S_i "simultaneously." Immediate philosophical difficulties arise. What does it mean to "count" the number of elements of S_i? There is no definitive answer to this question. Only through experience does one develop an idea of what is meant by a "determination" of a counting function f(i). The counting function f(i) can be given in several standard ways:

The most satisfactory form of f(i) is a completely explicit dosed formula involving only well-known functions, and free from summation symbols. Only in rare cases will such a formula exist. As formulas for f(i) become more complicated, the willingness to accept them as "determinations" of f(i) decreases. Consider the following examples:

Example: For each $n \in \mathbb{N}$, let f(n) be the number of subsets of the set [nJ = {1,2,...,n}. Then f(n) = 2^n, and no one will quarrel about this being a satisfactory formula for f(n).

Example: Suppose n men give their n hats to a hat-check person. Let f(n) be the number of ways that the hats can be given back to the men, each man receiving one hat, so that no man receives his own hat. For instance, f(1) = 0, f(2) = 1, f(3) = 2.

$$f(n) = n! \sum_{i=0}^{n} (-1)^i / i!.$$

This formula for f(n) is not elegant, but for lack of a simpler answer we are willing to accept $f(n) = n! \sum_{i=0}^{n} (-1)^i / i!$.as a satisfactory formula. In fact, once the derivation of $f(n) = n! \sum_{i=0}^{n} (-1)^i / i!$.is understood (using the Principle of InclusionExclusion), every term of $f(n) = n!\sum_{i=0}^{n}(-1)^i/i!$.has an easily understood combinatorial meaning. This enables us to "understand" $f(n) = n!\sum_{i=0}^{n}(-1)^i/i!$.intuitively, so our willingness to accept it is enhanced. We also remark that it follows easily from $f(n) = n!\sum_{i=0}^{n}(-1)^i/i!$. that f(n) is the nearest integer to $n!/e$. This is certainly a simple explicit formula, but it has the disadvantage of being "non-combinatorial"; that is, dividing by e and rounding off to the nearest integer has no direct combinatorial significance.

Example: Let f(n) be the number of n x n matrices **M** of zeros and ones such that every row and column of **M** has three ones. For example, f(O) = f(l) = f(2) = 0, f(3) = 1. The most explicit formula known at present for f(n) is,

$$f(n)=6^{-n}\sum\frac{(-1)^{\beta}n!^{2}(\beta+3y)!2^{\alpha}3^{\beta}}{\alpha!\beta!y!^{2}6^{y}}$$

where the sum is over all (n + 2)(n + 1)/2 solutions to $\alpha + ß + Y = n$ in nonnegative integers. This formula gives very little insight into the behavior of f(n), but it does allow one to compute f(n) much faster than if only the combinatorial definition of f(n) were used. Hence with some reluctance we accept $f(n)=6^{-n}\sum\frac{(-1)^{\beta}n!^{2}(\beta+3y)!2^{\alpha}3^{\beta}}{\alpha!\beta!y!^{2}6^{y}}$ as a "determination" of f(n). Of course if someone were later to prove f(n) = n(n - l)(n - 2)/6 (rather unlikely), then our enthusiasm for $f(n)=6^{-n}\sum\frac{(-1)^{\beta}n!^{2}(\beta+3y)!2^{\alpha}3^{\beta}}{\alpha!\beta!y!^{2}6^{y}}$ would be considerably diminished.

Example: There are actually formulas in the literature ("nameless here for evermore") for certain counting functions f(n) whose evaluation requires listing all (or almost all) of the f(n) objects being counted! Such a "formula" is completely worthless.

A recurrence for f(i) may be given in terms of previously calculated f(j)'s, thereby giving a simple procedure for calculating f(i) for any desired $i \in 1$. For instance, let f(n) be the number of subsets of [n] that do not contain two consecutive integers. For example, for n = 4 we have the subsets $\varnothing$, {I}, {2}, {3}, {4}, {l,3}, {1,4}, {2,4}, so f(4) = 8. It is easily seen that f(n) = f(n - 1) + f(n - 2) for n ~ 2. This makes it trivial, for example, to compute f(20). On the other hand, it can be shown that,

$$f(n)=\frac{1}{\sqrt{5}}\left(\tau^{n+2}-\overline{\tau}^{n+2}\right),$$

where $\frac{1}{2}\left(1+\sqrt{5}\right),\overline{\tau}=\frac{1}{2}\left(1-\sqrt{5}\right)$ This is an explicit answer, but because it involves irrational numbers it is a matter of opinion whether it is a better answer than the recurrence f(n) = f(n - 1) + f(n - 2).

An estimate may be given for f(i). If $1 = \mathbb{N}$, this estimate frequently takes the form of an asymptotic formula f(n) ~ g(n), where g(n) is a "familiar function." The notation f(n) ~ g(n) means that $\lim_{n\to\infty} f(n)/g(n) = 1$. For instance, let f(n) be the function of example above. It can be shown that,

$$f(n) \sim e^{-2} 36^{-n} (3n)!.$$

For many purposes this estimate is superior to the "explicit" formula.

GENERATING FUNCTION

In mathematics, a generating function is a way of encoding an infinite sequence of numbers (a_n) by treating them as the coefficients of a power series. This formal power series is the generating function. Unlike an ordinary series, this formal series is allowed to diverge, meaning that the generating function is not always a true function and the "variable" is actually an indeterminate. Generating functions were first introduced by Abraham de Moivre in 1730, in order to solve the general linear recurrence problem. One can generalize to formal series in more than one indeterminate, to encode information about arrays of numbers indexed by several natural numbers.

There are various types of generating functions, including ordinary generating functions, exponential generating functions, Lambert series, Bell series, and Dirichlet series; definitions and examples are given below. Every sequence in principle has a generating function of each type (except that Lambert and Dirichlet series require indices to start at 1 rather than 0), but the ease with which they can be handled may differ considerably. The particular generating function, if any, that is most useful in a given context will depend upon the nature of the sequence and the details of the problem being addressed.

Generating functions are often expressed in closed form (rather than as a series), by some expression involving operations defined for formal series. These expressions in terms of the indeterminate x may involve arithmetic operations, differentiation with respect to x and composition with (i.e., substitution into) other generating functions; since these operations are also defined for functions, the result looks like a function of x. Indeed, the closed form expression can often be interpreted as a function that can be evaluated at (sufficiently small) concrete values of x, and which has the formal series as its series expansion; this explains the designation "generating functions". However such interpretation is not required to be possible, because formal series are not required to give a convergent series when a nonzero numeric value is substituted for x. Also, not all expressions that are meaningful as functions of x are meaningful as expressions designating formal series; for example, negative and fractional powers of x are examples of functions that do not have a corresponding formal power series.

Generating functions are not functions in the formal sense of a mapping from a domain to a codomain. Generating functions are sometimes called generating series, in that a series of terms can be said to be the generator of its sequence of term coefficients.

Ordinary Generating Function (OGF)

The ordinary generating function of a sequence a_n is,

$$G(a_n;x)=\sum_{n=0}^{\infty}a_n x^n.$$

When the term generating function is used without qualification, it is usually taken to mean an ordinary generating function.

If a_n is the probability mass function of a discrete random variable, then its ordinary generating function is called a probability-generating function.

The ordinary generating function can be generalized to arrays with multiple indices. For example, the ordinary generating function of a two-dimensional array $a_{m,n}$ (where n and m are natural numbers) is,

$$G(a_{m,n};x,y)=\sum_{m,n=0}^{\infty}a_{m,n}x^m y^n.$$

Exponential Generating Function (EGF)

The exponential generating function of a sequence a_n is,

$$EG(a_n;x)=\sum_{n=0}^{\infty}a_n\frac{x^n}{n!}.$$

Exponential generating functions are generally more convenient than ordinary generating functions for combinatorial enumeration problems that involve labelled objects.

Poisson Generating Function

The Poisson generating function of a sequence a_n is,

$$PG(a_n;x)=\sum_{n=0}^{\infty}a_n e^{-x}\frac{x^n}{n!}=e^{-x}EG(a_n;x).$$

Lambert Series

The Lambert series of a sequence a_n is,

$$LG(a_n;x)=\sum_{n=1}^{\infty}a_n\frac{x^n}{1-x^n}.$$

The Lambert series coefficients in the power series expansions $b_n := [x^n]LG(a_n;x)$ for integers $n \geq 1$ are related by the divisor sum $b_n = \sum_{d|n} a_d$. In a Lambert series the index n starts at 1, not at 0, as the first term would otherwise be undefined.

Bell Series

The Bell series of a sequence a_n is an expression in terms of both an indeterminate x and a prime p and is given by,

$$BG_p(a_n;x) = \sum_{n=0}^{\infty} a_{p^n} x^n.$$

Dirichlet Series Generating Functions (DGFs)

Formal Dirichlet series are often classified as generating functions, although they are not strictly formal power series. The Dirichlet series generating function of a sequence a_n is,

$$DG(a_n;s) = \sum_{n=1}^{\infty} \frac{a_n}{n^s}.$$

The Dirichlet series generating function is especially useful when a_n is a multiplicative function, in which case it has an Euler product expression in terms of the function's Bell series,

$$DG(a_n;s) = \prod_p BG_p(a_n;p^{-s}).$$

If a_n is a Dirichlet character then its Dirichlet series generating function is called a Dirichlet L-series. We also have a relation between the pair of coefficients in the Lambert series expansions above and their DGFs. Namely, we can prove that $[x^n]LG(a_n;x) = b_n$ if and only if $DG(a_n;s)\zeta(s) = DG(b_n;s)$ where $\zeta(s)$ is the Riemann zeta function.

Polynomial Sequence Generating Functions

The idea of generating functions can be extended to sequences of other objects. Thus, for example, polynomial sequences of binomial type are generated by,

$$e^{xf(t)} = \sum_{n=0}^{\infty} \frac{p_n(x)}{n!} t^n$$

where $p_n(x)$ is a sequence of polynomials and f(t) is a function of a certain form. Sheffer sequences are generated in a similar way.

Ordinary Generating Functions

Examples of Generating Functions for Simple Sequences

Polynomials are a special case of ordinary generating functions, corresponding to finite sequences, or equivalently sequences that vanish after a certain point. These are important in that many finite sequences can usefully be interpreted as generating functions, such as the Poincaré polynomial and others.

A key generating function is that of the constant sequence 1, 1, 1, 1, 1, 1, 1, 1, 1, whose ordinary generating function is the geometric series:

$$\sum_{n=0}^{\infty} x^n = \frac{1}{1-x}.$$

The left-hand side is the Maclaurin series expansion of the right-hand side. Alternatively, the equality can be justified by multiplying the power series on the left by $1 - x$, and checking that the result is the constant power series 1 (in other words, that all coefficients except the one of x^0 are equal to 0). Moreover, there can be no other power series with this property. The left-hand side therefore designates the multiplicative inverse of $1 - x$ in the ring of power series.

Expressions for the ordinary generating function of other sequences are easily derived from this one. For instance, the substitution $x \to ax$ gives the generating function for the geometric sequence 1, a, a^2, a^3, for any constant a:

$$\sum_{n=0}^{\infty} (ax)^n = \frac{1}{1-ax}.$$

(The equality also follows directly from the fact that the left-hand side is the Maclaurin series expansion of the right-hand side.) In particular,

$$\sum_{n=0}^{\infty} (-1)^n x^n = \frac{1}{1+x}.$$

One can also introduce regular "gaps" in the sequence by replacing x by some power of x, so for instance for the sequence 1, 0, 1, 0, 1, 0, 1, 0, one gets the generating function,

$$\sum_{n=0}^{\infty} x^{2n} = \frac{1}{1-x^2}.$$

By squaring the initial generating function, or by finding the derivative of both sides with respect to x and making a change of running variable $n \to n + 1$, one sees that the coefficients form the sequence 1, 2, 3, 4, 5, ..., so one has,

$$\sum_{n=0}^{\infty} (n+1)x^n = \frac{1}{(1-x)^2},$$

and the third power has as coefficients the triangular numbers 1, 3, 6, 10, 15, 21, ... whose term n is the binomial coefficient $\binom{n+2}{2}$, so that,

$$\sum_{n=0}^{\infty}\binom{n+2}{2}x^n=\frac{1}{(1-x)^3}.$$

More generally, for any non-negative integer k and non-zero real value a, it is true that,

$$\sum_{n=0}^{\infty}a^n\binom{n+k}{k}x^n=\frac{1}{(1-ax)^{k+1}}.$$

Since,

$$2\binom{n+2}{2}-3\binom{n+1}{1}+\binom{n}{0}=2\frac{(n+1)(n+2)}{2}-3(n+1)+1=n^2,$$

one can find the ordinary generating function for the sequence 0, 1, 4, 9, 16, of square numbers by linear combination of binomial-coefficient generating sequences:

$$G(n^2;x)=\sum_{n=0}^{\infty}n^2x^n=\frac{2}{(1-x)^3}-\frac{3}{(1-x)^2}+\frac{1}{1-x}=\frac{x(x+1)}{(1-x)^3}.$$

We may also expand alternately to generate this same sequence of squares as a sum of derivatives of the geometric series in the following form:

$$\begin{aligned}G(n^2;x)&=\sum_{n=0}^{\infty}n^2x^n=\sum_{n=0}^{\infty}n(n-1)x^n+\sum_{n=0}^{\infty}nx^n\\&=x^2D^2\left[\frac{1}{1-x}\right]+xD\left[\frac{1}{1-x}\right]\\&=\frac{2x^2}{(1-x)^3}+\frac{x}{(1-x)^2}=\frac{x(x+1)}{(1-x)^3}.\end{aligned}$$

By induction, we can similarly show for positive integers $m\geq 1$ that:

$$n^m=\sum_{j=0}^{m}\left\{\begin{matrix}m\\j\end{matrix}\right\}\frac{n!}{(n-j)!},$$

where $\left\{\begin{matrix}n\\k\end{matrix}\right\}$ denote the Stirling numbers of the second kind and where the generating function $\sum_{n\geq 0}n!/(n-j)!z^n=j!\cdot z^j/(1-z)^{j+1}$, so that we can form the analogous generating functions over the integral m-th powers generalizing the result in the square case above. In particular, since we can write $\frac{z^k}{(1-z)^{k+1}}=\sum_{i=0}^{k}\binom{k}{i}\frac{(-1)^{k-i}}{(1-z)^{i+1}}$, we can apply a well-known finite sum identity involving the Stirling numbers to obtain that,

$$\sum_{n\geq 0} n^m z^n = \sum_{j=0}^{m} \left\{ {m+1 \atop j+1} \right\} \frac{(-1)^{m-j} j!}{(1-z)^{j+1}}.$$

Rational Functions

The ordinary generating function of a sequence can be expressed as a rational function (the ratio of two finite-degree polynomials) if and only if the sequence is a linear recursive sequence with constant coefficients; this generalizes the examples above. Conversely, every sequence generated by a fraction of polynomials satisfies a linear recurrence with constant coefficients; these coefficients are identical to the coefficients of the fraction denominator polynomial (so they can be directly read off). This observation shows it is easy to solve for generating functions of sequences defined by a linear finite difference equation with constant coefficients, and then hence, for explicit closed-form formulas for the coefficients of these generating functions. The prototypical example here is to derive Binet's formula for the Fibonacci numbers via generating function techniques.

We also notice that the class of rational generating functions precisely corresponds to the generating functions that enumerate quasi-polynomial sequences of the form,

$$f_n = p_1(n)\rho_1^n + \cdots + p_\ell(n)\rho_\ell^n,$$

where the reciprocal roots, $\rho_i \in \mathbb{C}$, are fixed scalars and where $p_i(n)$ is a polynomial in for all $1 \leq i \leq \ell$.

In general, Hadamard products of rational functions produce rational generating functions. Similarly, if $F(s,t) := \sum_{m,n\geq 0} f(m,n) w^m z^n$ is a bivariate rational generating function, then its corresponding diagonal generating function, $\mathrm{diag}(F) := \sum_{n\geq 0} f(n,n) z^n$, is algebraic. For example, if we let,

$$F(s,t) := \sum_{i,j\geq 0} \binom{i+j}{i} s^i t^j = \frac{1}{1-s-t},$$

then this generating function's diagonal coefficient generating function is given by the well-known OGF formula,

$$\mathrm{diag}(F) = \sum_{n\geq 0} \binom{2n}{n} z^n = \frac{1}{\sqrt{1-4z}}.$$

This result is computed in many ways, including Cauchy's integral formula or contour integration, taking complex residues, or by direct manipulations of formal power series in two variables.

Operations on Generating Functions

Multiplication Yields Convolution

Multiplication of ordinary generating functions yields a discrete convolution (the Cauchy product) of the sequences. For example, the sequence of cumulative sums (compare to the slightly more general Euler–Maclaurin formula),

$$(a_0, a_0 + a_1, a_0 + a_1 + a_2, \ldots)$$

of a sequence with ordinary generating function $G(a_n; x)$ has the generating function,

$$G(a_n; x) \cdot \frac{1}{1-x}$$

because $1/(1 - x)$ is the ordinary generating function for the sequence (1, 1,...).

Shifting Sequence Indices

For integers $m \geq 1$, we have the following two analogous identities for the modified generating functions enumerating the shifted sequence variants of $\langle g_{n-m} \rangle$ and $\langle g_{n+m} \rangle$, respectively:

$$z^m G(z) = \sum_{n \geq m} g_{n-m} z^n$$

$$\frac{G(z) - g_0 - g_1 z - \cdots - g_{m-1} z^{m-1}}{z^m} = \sum_{n \geq 0} g_{n+m} z^n.$$

Differentiation and Integration of Generating Functions

We have the following respective power series expansions for the first derivative of a generating function and its integral:

$$G'(z) = \sum_{n \geq 0} (n+1) g_{n+1} z^n$$

$$z \cdot G'(z) = \sum_{n \geq 0} n g_n z^n$$

$$\int_0^z G(t)dt = \sum_{n \geq 1} \frac{g_{n-1}}{n} z^n.$$

The differentiation–multiplication operation of the second identity can be repeated *k* times to multiply the sequence by n^k, but that requires alternating between differentiation and multiplication. If instead doing *k* differentiations in sequence, the effect is to multiply by the kth falling factorial:

$$z^k G^{(k)}(z) = \sum_{n \geq 0} n^{\underline{k}} g_n z^n = \sum_{n \geq 0} n(n-1)\cdots(n-k+1) g_n z^n \text{ for all } k \in \mathbb{N}.$$

Using the Stirling numbers of the second kind, that can be turned into another formula for multiplying by n^k as follows:

$$\sum_{j=0}^{k}\left\{{k \atop j}\right\} z^j F^{(j)}(z) = \sum_{n\geq 0} n^k f_n z^n \text{ for all } k \in \mathbb{N}.$$

A negative-order reversal of this sequence powers formula corresponding to the operation of repeated integration is defined by the zeta series transformation and its generalizations defined as a derivative-based transformation of generating functions, or alternately termwise by an performing an integral transformation on the sequence generating function.

Enumerating Arithmetic Progressions of Sequences

We give formulas for generating functions enumerating the sequence $\{f_{an+b}\}$ given an ordinary generating function $F(z)$ where $a, b \in \mathbb{N}$, $a \geq 2$, and $0 \leq b < a$ e main article on transformations). For $a = 2$, this is simply the familiar decomposition of a function into even and odd parts (i.e., even and odd powers):

$$\sum_{n\geq 0} f_{2n} z^{2n} = \frac{1}{2}\left(F(z) + F(-z)\right)$$

$$\sum_{n\geq 0} f_{2n+1} z^{2n+1} = \frac{1}{2}\left(F(z) - F(-z)\right).$$

More generally, suppose that $a \geq 3$ and that $\omega_a = \exp\left(2\pi\imath / a\right)$ denotes the ath primitive root of unity. Then, as an application of the discrete Fourier transform, we have the formula,

$$\sum_{n\geq 0} f_{an+b} z^{an+b} = \frac{1}{a}\sum_{m=0}^{a-1} \omega_a^{-mb} F(\omega_a^m z).$$

For integers $m \geq 1$, another useful formula providing somewhat reversed floored arithmetic progressions — effectively repeating each coefficient m times — are generated by the identity,

$$\sum_{n\geq 0} f_{\left\lfloor \frac{n}{m} \right\rfloor} z^n = \frac{1-z^m}{1-z} F(z^m) = \left(1 + z + \cdots + z^{m-2} + z^{m-1}\right) F(z^m).$$

P-recursive Sequences and Holonomic Generating Functions

A formal power series (or function) $F(z)$ is said to be holonomic if it satisfies a linear differential equation of the form,

$$c_0(z)F^{(r)}(z) + c_1(z)F^{(r-1)}(z) + \cdots + c_r(z)F(z) = 0,$$

where the coefficients $c_i(z)$ are in the field of rational functions, $\mathbb{C}(z)$. Equivalently, $F(z)$ is holonomic if the vector space over $\mathbb{C}(z)$ spanned by the set of all of its derivatives is finite dimensional.

Since we can clear denominators if need be in the previous equation, we may assume that the functions, $c_i(z)$ are polynomials in z. Thus we can see an equivalent condition that a generating function is holonomic if its coefficients satisfy a P-recurrence of the form,

$$\hat{c}_s(n)f_{n+s} + \hat{c}_{s-1}(n)f_{n+s-1} + \cdots + \hat{c}_0(n)f_n = 0,$$

for all large enough $n \geq n_0$ and where the $\hat{c}_i(n)$ are fixed finite-degree polynomials in n. In other words, the properties that a sequence be P-recursive and have a holonomic generating function are equivalent. Holonomic functions are closed under the Hadamard product operation $\odot$ on generating functions.

Examples:

The functions e^z, $\log(z)$, $\cos(z)$, $\arcsin(z)$, $\sqrt{1+z}$, the dilogarithm function $\mathrm{Li}_2(z)$, the generalized hypergeometric functions ${}_pF_q(\ldots;\ldots;z)$ and the functions defined by the power series $\sum_{\geq} z \ /(n!)$ and the non-convergent $\sum_{n\geq 0} n! \cdot z^n$ are all holonomic. Examples of P-recursive sequences with holonomic generating functions include $f_n := \frac{1}{n+1}\binom{2n}{n}$ and $f_n := 2^n/(n^2+1)$, where sequences such as $\sqrt{n}$ and $\log(n)$ are not P-recursive due to the nature of singularities in their corresponding generating functions. Similarly, functions with infinitely-many singularities such as $\tan(z)$, $\sec(z)$, and $\Gamma(z)$ are not holonomic functions.

Software for Working with P-recursive Sequences and Holonomic Generating Functions

Tools for processing and working with P-recursive sequences in Mathematica include the software packages provided for non-commercial use on the RISC Combinatorics Group algorithmic combinatorics software site. Despite being mostly closed-source, particularly powerful tools in this software suite are provided by the Guess package for guessing P-recurrences for arbitrary input sequences (useful for experimental mathematics and exploration) and the Sigma package which is able to find P-recurrences for many sums and solve for closed-form solutions to P-recurrences involving generalized harmonic numbers. Other packages listed on this particular RISC site are targeted at working with holonomic generating functions specifically.

Relation to Discrete-time Fourier Transform

When the series converges absolutely,

$$G\left(a_n;e^{-i\omega}\right)=\sum_{n=0}^{\infty}a_n e^{-i\omega n}$$

is the discrete-time Fourier transform of the sequence a_0, a_1, ...

Asymptotic Growth of a Sequence

In calculus, often the growth rate of the coefficients of a power series can be used to deduce a radius of convergence for the power series. The reverse can also hold; often the radius of convergence for a generating function can be used to deduce the asymptotic growth of the underlying sequence.

For instance, if an ordinary generating function $G(a_n; x)$ that has a finite radius of convergence of r can be written as,

$$(a_n;x)=\frac{A(x)+B(x)\left(1-\frac{x}{r}\right)^{-\beta}}{x^{\alpha}}$$

where each of A(x) and B(x) is a function that is analytic to a radius of convergence greater than r (or is entire), and where $B(r) \neq 0$ then,

$$a_n \sim \frac{B(r)}{r^{\alpha}\Gamma(\beta)}n^{\beta-1}(1/r)^n \sim \frac{B(r)}{r^{\alpha}}\binom{n+\beta-1}{n}(1/r)^n = \frac{B(r)}{r^{\alpha}}\left(\!\binom{\beta}{n}\!\right)(1/r)^n,$$

using the Gamma function, a binomial coefficient, or a multiset coefficient.

Often this approach can be iterated to generate several terms in an asymptotic series for a_n. In particular,

$$G\left(a_n-\frac{B(r)}{r^{\alpha}}\binom{n+\beta-1}{n}(1/r)^n;x\right)=G(a_n;x)-\frac{B(r)}{r^{\alpha}}\left(1-\frac{x}{r}\right)^{-\beta}.$$

The asymptotic growth of the coefficients of this generating function can then be sought via the finding of A, B, α, β, and r to describe the generating function.

Similar asymptotic analysis is possible for exponential generating functions. With an exponential generating function, it is $a_n/n!$ that grows according to these asymptotic formulae.

Asymptotic Growth of the Sequence of Squares

The ordinary generating function for the sequence of squares is,

$$\frac{x(x+1)}{(1-x)^3}.$$

With r = 1, α = 0, β = 3, A(x) = 0, and B(x) = x(x+1), we can verify that the squares grow as expected, like the squares:

$$a_n \sim \frac{B(r)}{r^\alpha \Gamma(\beta)} n^{\beta-1} \left(\frac{1}{r}\right)^n = \frac{1(1+1)}{1^0 \Gamma(3)} n^{3-1} (1/1)^n = n^2.$$

Asymptotic Growth of the Catalan Numbers

The ordinary generating function for the Catalan numbers is,

$$\frac{1-\sqrt{1-4x}}{2x}.$$

With r = 1/4, α = 1, β = −1/2, A(x) = 1/2, and B(x) = −1/2, we can conclude that, for the Catalan numbers,

$$a_n \sim \frac{B(r)}{r^\alpha \Gamma(\beta)} n^{\beta-1} \left(\frac{1}{r}\right)^n = \frac{-\frac{1}{2}}{(\frac{1}{4})^1 \Gamma(-\frac{1}{2})} n^{-\frac{1}{2}-1} \left(\frac{1}{\frac{1}{4}}\right)^n = \frac{1}{\sqrt{\pi}} n^{-\frac{3}{2}} 4^n.$$

Bivariate and Multivariate Generating Functions

One can define generating functions in several variables for arrays with several indices. These are called multivariate generating functions or, sometimes, super generating functions. For two variables, these are often called bivariate generating functions.

For instance, since $(1+x)^n$ is the ordinary generating function for binomial coefficients for a fixed n, one may ask for a bivariate generating function that generates the binomial coefficients $\binom{n}{k}$ for all k and n. To do this, consider $(1+x)^n$ as itself a series, in n, and find thegenerating function in y that has these as coefficients. Since the generating function for a^n is,

$$\frac{\quad}{ay}$$

the generating function for the binomial coefficients is:

$$\sum_{n,k} \binom{n}{k} x^k y^n = \frac{1}{1-(1+x)y} = \frac{1}{1-y-xy}.$$

Representation by Continued Fractions (Jacobi-type J-Fractions)

Expansions of (formal) Jacobi-type and Stieltjes-type continued fractions (J-fractions and S-fractions, respectively) whose h^{th} rational convergents represent $2h$-order accurate power series are another way to express the typically divergent ordinary generating functions for many special one and two-variate sequences. The particular form of the Jacobi-type continued fractions (J-fractions) are expanded as in the following equation and have the next corresponding power series expansions with respect to z for some specific, application-dependent component sequences, $\{\mathrm{ab}_i\}$ and $\{c_i\}$, where $z \neq 0$ denotes the formal variable in the second power series expansion given below:

$$J^{[\infty]}(z) = \cfrac{1}{1 - c_1 z - \cfrac{ab_2 z^2}{1 - c_2 z - \cfrac{ab_3 z^2}{\ddots}}} = 1 + c_1 z + \left(ab_2 + c_1^2\right)z^2 + \left(2ab_2 c_1 + c_1^3 + ab_2 c_2\right)z^3 + \cdots.$$

The coefficients of z^n, denoted in shorthand by $j_n := [z^n]J^{[\infty]}(z)$, in the previous equations correspond to matrix solutions of the equations,

$$\begin{bmatrix} k_{0,1} & k_{1,1} & 0 & 0 & \cdots \\ k_{0,2} & k_{1,2} & k_{2,2} & 0 & \cdots \\ k_{0,3} & k_{1,3} & k_{2,3} & k_{3,3} & \cdots \\ \vdots & \vdots & \vdots & \vdots & \end{bmatrix} = \begin{bmatrix} k_{0,0} & 0 & 0 & 0 & \cdots \\ k_{0,1} & k_{1,1} & 0 & 0 & \cdots \\ k_{0,2} & k_{1,2} & k_{2,2} & 0 & \cdots \\ \vdots & \vdots & \vdots & \vdots & \end{bmatrix} \cdot \begin{bmatrix} c_1 & 1 & 0 & 0 & \cdots \\ ab_2 & c_2 & 1 & 0 & \cdots \\ 0 & ab_3 & c_3 & 1 & \cdots \\ \vdots & \vdots & \vdots & \vdots & \end{bmatrix},$$

where $j_0 \equiv k_{0,0} = 1$, $j_n = k_{0,n}$ for, $n \geq 1$ if $k_{r,s} = 0$, and where for all integers $r > s$, we have an addition formula relation given by,

$$j_{p+q} = k_{0,p} \cdot k_{0,q} + \sum_{i=1}^{\min(p,q)} ab_2 \cdots ab_{i+1} \times k_{i,p} \cdot k_{i,q}.$$

Properties of the hth Convergent Functions

For $h \geq 0$ (though in practice when $h \geq 2$), we can define the rational h^{th} convergents to the infinite J-fraction, $J^{[\infty]}(z)$, expanded by,

$$\mathrm{Conv}_h(z) := \frac{P_h(z)}{Q_h(z)} = j_0 + j_1 z + \cdots + j_{2h-1} z^{2h-1} + \sum_{n \geq 2h} \vec{j}_{h,n} z^n,$$

component-wise through the sequences, $P_h(z)$ and $Q_h(z)$, defined recursively by,

$$P_h(z) = (1 - c_h z)P_{h-1}(z) - ab_h z^2 P_{h-2}(z) + \delta_{h,1}$$
$$Q_h(z) = (1 - c_h z)Q_{h-1}(z) - ab_h z^2 Q_{h-2}(z) + (1 - c_1 z)\delta_{h,1} + \delta_{0,1}.$$

Moreover, the rationality of the convergent function, $\mathrm{Conv}_h(z)$ for all $h \geq 2$ implies additional finite difference equations and congruence properties satisfied by the sequence of j_n, and for $M_h := \mathrm{ab}_2 \cdots \mathrm{ab}_{h+1}$ if $h \,||\, M_h$ then we have the congruence $\mathrm{j}_n \equiv [\mathrm{z}^n]\mathrm{Conv}_h(\mathrm{z}) \pmod{\mathrm{h}}$, for non-symbolic, determinate choices of the parameter sequences, $\{\mathrm{ab}_i\}$ and $\{\mathrm{c}_i\}$, when $\mathrm{h} \geq 2$, i.e., when these sequences do not implicitly depend on an auxiliary parameter such as q, x, or R as in the examples contained in the table below.

Examples:

The next table provides examples of closed-form formulas for the component sequences found computationally (and subsequently proved correct in the cited references) in several special cases of the prescribed sequences, j_n, generated by the general expansions of the J-fractions defined in the first subsection. Here we define $0 < |a|, |b|, |q| < 1$ and the parameters R, $\alpha \in \mathbb{Z}^+$ and x to be indeterminates with respect to these expansions, where the prescribed sequences enumerated by the expansions of these J-fractions are defined in terms of the q-Pochhammer symbol, Pochhammer symbol, and the binomial coefficients.

j_n	c_1	$\mathrm{c}_i (i \geq 2)$	$\mathrm{ab}_i (i \geq 2)$
q^{n^2}	q	$q^{2h-3}\left(q^{2h} + q^{2h-2} - 1\right)$	$q^{6h-10}\left(q^{2h-2} - 1\right)$
$(a;q)_n$	$1-a$	$q^{h-1} - aq^{h-2}\left(q^h + q^{h-1} - 1\right)$	$aq^{2h-4}(aq^{h-2} - 1)(q^{h-1} - 1)$
$\left(zq^{-n};q\right)_n$	$\frac{q-z}{q}$	$\frac{q^h - z - qz + q^h z}{q^{2h-1}}$	$\frac{(q^{h-1}-1)(q^{h-1}-z)\cdot z}{q^{4h-5}}$
$\frac{(a;q)}{(b;q)}$	$\frac{1-a}{1-b}$	$\frac{q^{i-2}\left(q + abq^{2i-3} + a(1-q^{i-1}-q^i) + b(q^i - q - 1)\right)}{(1-bq^{2i-4})(1-bq^{2i-2})}$	$\frac{q^{2i-4}(1-bq^{i-3})(1-aq^{i-2})(a-bq^{i-2})(1-q^{i-1})}{(1-bq^{2i-5})(1-bq^{2i-4})^2(1-bq^{2i-3})}$
$\alpha^n \cdot \left(\frac{R}{\alpha}\right)_n$	R	$R + 2\alpha(i-1)$	$(i-1)\alpha(R + (i-2)\alpha)$
$(-1)^n \binom{x}{n}$	$-x$	$-\frac{(x + 2(i-1)^2)}{(2i-1)(2i-3)}$	$\begin{cases} -\frac{(x-i+2)(x+i-1)}{4\cdot(2i-3)^2} & i \geq 3; \\ -\frac{1}{2}x(x+1) & i = 2. \end{cases}$
$(-1)^n \binom{x+n}{n}$	$-(x+1)$	$\frac{(x - 2i(i-2) - 1)}{(2i-1)(2i-3)}$	$\begin{cases} -\frac{(x-i+2)(x+i-1)}{4\cdot(2i-3)^2} & i \geq 3; \\ -\frac{1}{2}x(x+1) & i = 2. \end{cases}$

The radii of convergence of these series corresponding to the definition of the Jacobi-type J-fractions given above are in general different from that of the corresponding power series expansions defining the ordinary generating functions of these sequences.

Examples:

Generating functions for the sequence of square numbers $a_n = n^2$ are:

Ordinary Generating Function

$$G(n^2;x)=\sum_{n=0}^{\infty} n^2x^n = \frac{x(x+1)}{(1-x)^3}$$

Exponential Generating Function

$$EG(n^2;x)=\sum_{n=0}^{\infty}\frac{n^2x^n}{n!}=x(x+1)e^x$$

Lambert Series

We can show that for $|x|,|xq|<1$ we have that,

$$\sum_{n\geq 1}\frac{q^n x^n}{1-x^n}=\sum_{n\geq 1}\frac{q^n x^{n^2}}{1-qx^n}+\sum_{n\geq 1}\frac{q^n x^{n(n+1)}}{1-x^n},$$

where we have the special case identity for the generating function of the divisor function, $d(n)\equiv\sigma_0(n)$, given by,

$$\sum_{n\geq 1}\frac{x^n}{1-x^n}=\sum_{n\geq 1}\frac{x^{n^2}(1+x^n)}{1-x^n}.$$

Bell Series

$$BG_p(n^2;x)=\sum_{n=0}^{\infty}(p^n)^2x^n=\frac{1}{1-p^2x}$$

Dirichlet Series Generating Function

$$DG(n^2;s)=\sum_{n=1}^{\infty}\frac{n^2}{n^s}=\zeta(s-2),$$

using the Riemann zeta function.

The sequence a_k generated by a Dirichlet series generating function (DGF) corresponding to:

$$DG(a_k;s)=\zeta(s)^m$$

where $\zeta(s)$ is the Riemann zeta function, has the ordinary generating function:

$$\sum_{k=1}^{k=n} a_k x^k = x + \binom{m}{1} \sum_{2\le a\le n} x^a + \binom{m}{2} \sum\sum_{\substack{a\ge 2\, b\ge 2 \\ ab\le n}} x^{ab} + \binom{m}{3} \sum\sum\sum_{\substack{a\ge 2\, c\ge 2\, b\ge 2 \\ abc\le n}} x^{abc} + \binom{m}{4} \sum\sum\sum\sum_{\substack{a\ge 2\, b\ge 2\, c\ge 2\, d\ge 2 \\ abcd\le n}} x^{abcd} + \cdots$$

Multivariate Generating Function

Multivariate generating functions arise in practice when calculating the number of contingency tables of non-negative integers with specified row and column totals. Suppose the table has r rows and c columns; the row sums are $t_1, \ldots t_r$ and the column sums are $s_1, \ldots s_c$. Then, according to I. J. Good, the number of such tables is the coefficient of,

$$x_1^{t_1} \ldots x_r^{t_r} y_1^{s_1} \ldots y_c^{s_c}$$

in,

$$\prod_{i=1}^{r} \prod_{j=1}^{c} \frac{1}{1 - x_i y_j}.$$

In the bivariate case, non-polynomial double sum examples of so-termed "double" or "super" generating functions of the form $G(w,z) := \sum_{m,n\ge 0} g_{m,n} w^m z^n$ include the following two-variable generating functions for the binomial coefficients, the Stirling numbers, and the Eulerian numbers:

$$e^{z+wz} = \sum_{m,n\ge 0} \binom{n}{m} w^m \frac{z^n}{n!}$$

$$e^{w(e^z-1)} = \sum_{m,n\ge 0} \left\{ {n \atop m} \right\} w^m \frac{z^n}{n!}$$

$$\frac{1}{(1-z)^w} = \sum_{m,n\ge 0} \left[{n \atop m} \right] w^m \frac{z^n}{n!}$$

$$\frac{1-w}{e^{(w-1)z} - w} = \sum_{m,n\ge 0} \left\langle {n \atop m} \right\rangle w^m \frac{z^n}{n!}$$

$$\frac{e^w - e^z}{we^z - ze^w} = \sum_{m,n\ge 0} \left\langle {m+n+1 \atop m} \right\rangle \frac{w^m z^n}{(m+n+1)!}.$$

Various Techniques: Evaluating Sums and Tackling other Problems with Generating Functions

Example: A Formula for Sums of Harmonic Numbers

Generating functions give us several methods to manipulate sums and to establish identities between sums.

The simplest case occurs when $s_n = \sum_{k=0}^{n} a_k$. We then know that $S(z) = \frac{A(z)}{1-z}$ for the corresponding ordinary generating functions.

For example, we can manipulate $s_n = \sum_{k=1}^{n} H_k$, where $H_k = 1 + \frac{1}{2} + \cdots + \frac{1}{k}$ are the harmonic numbers. Let $H(z) = \sum_{n\geq 1} H_n z^n$ be the ordinary generating function of the harmonic numbers. Then,

$$H(z) = \frac{\sum_{n\geq 1} \frac{1}{n} z^n}{1-z},$$

and thus,

$$S(z) = \sum_{n\geq 1} s_n z^n = \frac{\sum_{n\geq 1} \frac{1}{n} z^n}{(1-z)^2}.$$

Using $\frac{1}{(1-z)^2} = \sum_{n\geq 0} (n+1) z^n$, convolution with the numerator yields,

$$S_n = \sum_{k=1}^{n} \frac{1}{k}(n+1-k) = (n+1)H_n - n,$$

which can also be written as,

$$\sum_{k=1}^{n} H_k = (n+1)(H_{n+1} - 1).$$

Example: Modified Binomial Coefficient Sums and the Binomial Transform

As another example of using generating functions to relate sequences and manipulate sums, for an arbitrary sequence $\langle f_n \rangle$ we define the two sequences of sums,

$$s_n := \sum_{m=0}^{n} \binom{n}{m} f_m 3^{n-m}$$

$$\tilde{s}_n := \sum_{m=0}^{n} \binom{n}{m} (m+1)(m+2)(m+3) f_m 3^{n-m},$$

for all $n \geq 0$, and seek to express the second sums in terms of the first. We suggest an approach by generating functions.

First, we use the binomial transform to write the generating function for the first sum as,

$$S(z) = \frac{1}{(1-3z)} F\left(\frac{z}{1-3z}\right).$$

Since the generating function for the sequence $\langle (n+1)(n+2)(n+3) f_n \rangle$ is given by $6F(z) + 18zF'(z) + 9z^2 F''(z) + z^3 F^{(3)}(z)$, we may write the generating function for the second sum defined above in the form,

$$\tilde{S}(z) = \frac{6}{(1-3z)} F\left(\frac{z}{1-3z}\right) + \frac{18z}{(1-3z)^2} F'\left(\frac{z}{1-3z}\right) + \frac{9z^2}{(1-3z)^3} F''\left(\frac{z}{1-3z}\right) + \frac{z^3}{(1-3z)^4} F^{(3)}\left(\frac{z}{1-3z}\right).$$

In particular, we may write this modified sum generating function in the form of,

$$a(z) \cdot S(z) + b(z) \cdot zS'(z) + c(z) \cdot z^2 S''(z) + d(z) \cdot z^3 S^{(3)}(z),$$

for $a(z) = 6(1-3z)^3$, $b(z) = 18(1-3z)^3$, $c(z) = 9(1-3z)^3$, and $d(z) = (1-3z)^3$ where $(1-3z)^3 = 1 - 9z + 27z^2 - 27z^3$.

Finally, it follows that we may express the second sums through the first sums in the following form:

$$\begin{aligned}\tilde{s}_n &= [z^n] \left(\begin{array}{l} 6(1-3z)^3 \sum_{n \geq 0} s_n z^n + 18(1-3z)^3 \sum_{n \geq 0} n s_n z^n + 9(1-3z)^3 \sum_{n \geq 0} n(n-1) s_n z^n \\ \qquad + (1-3z)^3 \sum_{n \geq 0} n(n-1)(n-2) s_n z^n \end{array} \right) \\ &= (n+1)(n+2)(n+3) s_n - 9n(n+1)(n+2) s_{n-1} + 27(n-1)n(n+1) s_{n-2} \\ &\qquad - (n-2)(n-1)n s_{n-3}.\end{aligned}$$

Convolution (Cauchy Products)

A discrete convolution of the terms in two formal power series turns a product of generating functions into a generating function enumerating a convolved sum of the original sequence terms.

- Consider A(z) and B(z) are ordinary generating functions.

$$C(z) = A(z)B(z) \Leftrightarrow [z^n]C(z) = \sum_{k=0}^{n} a_k b_{n-k}$$

- Consider A(z) and B(z) are exponential generating functions.

$$C(z) = A(z)B(z) \Leftrightarrow [z^n / n!]C(z) = \sum_{k=0}^{n} \binom{n}{k} a_k b_{n-k}$$

- Consider the triply convolved sequence resulting from the product of three ordinary generating functions.

$$C(z) = F(z)G(z)H(z) \Leftrightarrow [z^n]C(z) = \sum_{j+k+\ell=n} f_j g_k h_\ell$$

- Consider the m-fold convolution of a sequence with itself for some positive integer $m \geq 1$.

$$C(z) = G(z)^m \Leftrightarrow [z^n]C(z) = \sum_{k_1+k_2+\cdots+k_m=n} g_{k_1} g_{k_2} \cdots g_{k_m}$$

Multiplication of generating functions, or convolution of their underlying sequences, can correspond to a notion of independent events in certain counting and probability scenarios. For example, if we adopt the notational convention that the probability generating function, or pgf, of a random variable Z is denoted by $G_Z(z)$, then we can show that for any two random variables,

$$G_{X+Y}(z) = G_X(z)G_Y(z),$$

if X and Y are independent. Similarly, the number of ways to pay $n \geq 0$ cents in coin denominations of values in the set $\{1,5,10,25,50\}$ (i.e., in pennies, nickels, dimes, quarters, and half dollars, respectively) is generated by the product,

$$C(z) = \frac{1}{1-z}\frac{1}{1-z^5}\frac{1}{1-z^{10}}\frac{1}{1-z^{25}}\frac{1}{1-z^{50}},$$

and moreover, if we allow the n cents to be paid in coins of any positive integer denomination, we arrive at the generating for the number of such combinations of change being generated by the partition function generating function expanded by the infinite q-Pochhammer symbol product of $\prod_{n\geq 1}(1-z^n)^{-1}$.

Example: The Generating Function for the Catalan Numbers

An example where convolutions of generating functions are useful allows us to solve for a specific closed-form function representing the ordinary generating function for the Catalan numbers, C_n. In particular, this sequence has the combinatorial interpretation as being the number of ways to insert parentheses into the product $x_0 \cdot x_1 \cdots x_n$ so that the order of multiplication is completely specified. For example, $C_2 = 2$ which corresponds to the two expressions $x_0 \cdot (x_1 \cdot x_2)$ and $(x_0 \cdot x_1) \cdot x_2$. It follows that the sequence satisfies a recurrence relation given by,

$$C_n = \sum_{k=0}^{n-1} C_k C_{n-1-k} + \delta_{n,0} = C_0 C_{n-1} + C_1 C_{n-2} + \cdots + C_{n-1} C_0 + \delta_{n,0}, n \geq 0,$$

and so has a corresponding convolved generating function, $C(z)$, satisfying,

$$C(z) = z \cdot C(z)^2 + 1.$$

Since $C(0) = 1 \neq \infty$, we then arrive at a formula for this generating function given by,

$$\begin{aligned} C(z) &= \frac{1 - \sqrt{1 - 4z}}{2z} \\ &= \sum_{n \geq 0} \frac{1}{n+1} \binom{2n}{n} z^n. \end{aligned}$$

Note that the first equation implicitly defining $C(z)$ above implies that,

$$C(z) = \frac{1}{1 - z \cdot C(z)},$$

which then leads to another "simple" (as in of form) continued fraction expansion of this generating function.

Example: Spanning Trees of Fans and Convolutions

A fan of order n is defined to be a graph on the vertices $\{0, 1, \ldots, n\}$ with $2n - 1$ edges connected according to the following rules: Vertex 0 is connected by a single edge to each of the other n vertices, and vertex k is connected by a single edge to the next vertex $k+1$ for all $1 \leq k < n$. There is one fan of order one, three fans of order two, eight fans of order three, and so on. A spanning tree is a subgraph of a graph which contains all of the original vertices and which contains enough edges to make this subgraph connected, but not so many edges that there is a cycle in the subgraph. We ask how many spanning trees f_n of a fan of order n are possible for each $n \geq 1$.

As an observation, we may approach the question by counting the number of ways to join adjacent sets of vertices. For example, when $n = 4$, we have that $f_4 = 4 + 3 \cdot 1 + 2 \cdot 2 + 1 \cdot 3 + 2 \cdot 1 \cdot 1 + 1 \cdot 2 \cdot 1 + 1 \cdot 1 \cdot 2 + 1 \cdot 1 \cdot 1 \cdot 1 \cdot 1 = 21$, which is a sum over the $g_n = n = [z^n] z / (1 - z)^2$-fold convolutions of the sequence $g_n = n = [z^n] z / (1 - z)^2$ for $m := 1, 2, 3, 4$. More generally, we may write a formula for this sequence as,

$$f_n = \sum_{m>0} \sum_{\substack{k_1+k_2+\cdots+k_m=n \\ k_1,k_2,\ldots,k_m>0}} g_{k_1} g_{k_2} \cdots g_{k_m},$$

from which we see that the ordinary generating function for this sequence is given by the next sum of convolutions as,

$$\begin{aligned} F(z) &= G(z) + G(z)^2 + G(z)^3 + \cdots \\ &= \frac{G(z)}{1 - G(z)} \\ &= \frac{z}{(1-z)^2 - z} \\ &= \frac{z}{1 - 3z + z^2}, \end{aligned}$$

from which we are able to extract an exact formula for the sequence by taking the partial fraction expansion of the last generating function.

Introducing a Free Parameter (Snake Oil Method)

Sometimes the sum s_n is complicated, and it is not always easy to evaluate. The "Free Parameter" method is another method (called "snake oil" by H. Wilf) to evaluate these sums.

Both methods discussed so far have *n* as limit in the summation. When n does not appear explicitly in the summation, we may consider *n* as a "free" parameter and treat s_n as a coefficient of $F(z) = \sum s_n z^n$, change the order of the summations on *n* and *k*, and try to compute the inner sum.

For example, if we want to compute,

$$s_n = \sum_{k \ge 0} \binom{n+k}{m+2k} \binom{2k}{k} \frac{(-1)^k}{k+1} \quad (m, n \in \mathbb{N}_0)$$

we can treat *n* as a "free" parameter, and set,

$$F(z) = \sum_{n \ge 0} \left[\sum_{k \ge 0} \binom{n+k}{m+2k} \binom{2k}{k} \frac{(-1)^k}{k+1} \right] z^n$$

Interchanging summation ("snake oil") gives,

$$F(z) = \sum_{k \ge 0} \binom{2k}{k} \frac{(-1)^k}{k+1} z^{-k} \sum_{n \ge 0} \binom{n+k}{m+2k} z^{n+k}$$

Now the inner sum is $\frac{z^{m+2k}}{(1-z)^{m+2k+1}}$. Thus,

$$F(z)=\frac{z^m}{(1-z)^{m+1}}\sum_{k\geq 0}\frac{1}{k+1}\binom{2k}{k}\left(\frac{-z}{(1-z)^2}\right)^k$$

$$=\frac{z^m}{(1-z)^{m+1}}\sum_{k\geq 0}C_k\left(\frac{-z}{(1-z)^2}\right)^k \quad \text{(where } C_k = k^{\text{th}} \text{ Catalan number)}$$

$$=\frac{z^m}{(1-z)^{m+1}}\frac{1-\sqrt{1+\frac{4z}{(1-z)^2}}}{\frac{-2z}{(1-z)^2}}$$

$$=\frac{-z^{m-1}}{2(1-z)^{m-1}}\left(1-\frac{1+z}{1-z}\right)$$

$$=\frac{z^m}{(1-z)^m}=z\frac{z^{m-1}}{(1-z)^m}.$$

Then we obtain,

$$s_n=\binom{n-1}{m-1} \text{ for } m\geq 1,\ s_n=[n=0] \text{ for } m=0.$$

Generating Functions Prove Congruences

We say that two generating functions (power series) are congruent modulo m, written $A(z)\equiv B(z)\pmod{m}$ if their coefficients are congruent modulo m for all $n\geq 0$, i.e., $a_n\equiv b_n\pmod{m}$ for all relevant cases of the integers n (note that we need not assume that m is an integer here—it may very well be polynomial-valued in some indeterminate x, for example). If the "simpler" right-hand-side generating function, $B(z)$, is a rational function of $m\geq 2$, then the form of this sequences suggests that the sequence is eventually periodic modulo fixed particular cases of integer-valued $m\geq 2$. For example, we can prove that the Euler numbers, $\langle E_n\rangle=\langle 1,1,5,61,1385,\ldots\rangle\mapsto\langle 1,1,2,1,2,1,2,\ldots\rangle \pmod{3}$, satisfy the following congruence modulo 3.

$$\sum_{n\geq 0}E_n z^n=\frac{1-z^2}{1+z^2}\pmod{3}.$$

One of the most useful, if not downright powerful, methods of obtaining congruences for sequences enumerated by special generating functions modulo any integers (i.e., not only prime powers p^k) is given on continued fraction representations of (even non-convergent) ordinary generating functions by J-fractions above. We cite one particular result related to generating series expanded through a representation by continued fraction from Lando's Lectures on Generating Functions as follows:

> Theorem: (Congruences for Series Generated by Expansions of Continued Fractions) Suppose that the generating function $A(z)$ is represented by an infinite continued fraction of the form,
>
> $$A(z) = \frac{1}{1-c_1 z} \frac{p_1 z^2}{1-c_2 z} \frac{p_2 z^2}{1-c_3 z} \cdots,$$
>
> and that $A_p(z)$ denotes the pth convergent to this continued fraction expansion defined such that $a_n = [z^n]A_p(z)$ for all $0 \le n < 2p$. Then 1) the function $A_p(z)$ is rational for all $p \ge 2$ where we assume that one of divisibility criteria of $p \mid p_1, p_1p_2, p_1p_2p_3 \cdots$ is met, i.e., $p \mid p_1 p_2 \cdots p_k$ for some $k \ge 1$; and 2) If the integer p divides the product $p_1p_2 \cdots p_k$, then we have that $A(z) \equiv A_k(z) \pmod{p}$.

Generating functions also have other uses in proving congruences for their coefficients. We cite the next two specific examples deriving special case congruences for the Stirling numbers of the first kind and for the partition function (mathematics) $p(n)$ which show the versatility of generating functions in tackling problems involving integer sequences.

The Stirling Numbers Modulo Small Integers

Stirling numbers generated by the finite products,

$$S_n(x) := \sum_{k=0}^{n} \begin{bmatrix} n \\ k \end{bmatrix} x^k = x(x+1)(x+2)\cdots(x+n-1), n \ge 1,$$

provides an overview of the congruences for these numbers derived strictly from properties of their generating function. We repeat the basic argument and notice that when reduces modulo 2, these finite product generating functions each satisfy,

$$S_n(x) = [x(x+1)]\cdot[x(x+1)]\cdots = x^{\lfloor n/2 \rfloor}(x+1)^{\lfloor n/2 \rfloor},$$

which implies that the parity of these Stirling numbers matches that of the binomial coefficient,

$$\begin{bmatrix} n \\ k \end{bmatrix} \equiv \binom{\lfloor n/2 \rfloor}{k - \lceil n/2 \rceil} \pmod{2},$$

and consequently shows that $\begin{bmatrix} n \\ k \end{bmatrix}$ is even whenever $k < \left\lceil \frac{n}{2} \right\rceil$.

Similarly, we can reduce the right-hand-side products defining the Stirling number generating functions modulo 3 to obtain slightly more complicated expressions providing that

$$\begin{bmatrix} n \\ m \end{bmatrix} \equiv [x^m]\left(x^{\lceil n/3 \rceil}(x+1)^{\lceil (n-1)/3 \rceil}(x+2)^{\lfloor n/3 \rfloor}\right) \pmod 3$$
$$\equiv \sum_{k=0}^{m} \binom{\lceil (n-1)/3 \rceil}{k} \binom{\lfloor n/3 \rfloor}{m-k-\lceil n/3 \rceil} \times 2^{\lfloor n/3 \rfloor + \lfloor n/3 \rfloor - (m-k)} \pmod 3.$$

Congruences for the Partition Function

In this example, we pull in some of the machinery of infinite products whose power series expansions generate the expansions of many special functions and enumerate partition functions. In particular, we recall that the partition function $p(n)$ is generated by the reciprocal infinite q-Pochhammer symbol product (or z-Pochhammer product as the case may be) given by,

$$\sum_{n \geq 1} p(n) z^n = \frac{1}{(1-z)(1-z^2)(1-z^3)\cdots}$$
$$= 1 + z + 2z^2 + 3z^3 + 5z^4 + 7z^5 + 11z^6 + \cdots.$$

This partition function satisfies many known congruence properties, which notably include the following results though there are still many open questions about the forms of related integer congruences for the function:

$$p(5m+4) \equiv 0 \pmod 5$$
$$p(7m+5) \equiv 0 \pmod 7$$
$$p(11m+6) \equiv 0 \pmod{11}$$
$$p(25m+24) \equiv 0 \pmod{5^2}.$$

We show how to use generating functions and manipulations of congruences for formal power series to give a highly elementary proof of the first of these congruences listed above.

First, we observe that the binomial coefficient generating function, $1/(1-z)^5$, satisfies that each of its coefficients are divisible by 5 with the exception of those which correspond to the powers of $1, z^5, z^{10}, \ldots$, all of which otherwise have a remainder of 1 modulo 5. Thus we may write,

$$\frac{1}{(1-z)^5} \equiv \frac{1}{1-z^5} \pmod 5 \quad \Leftrightarrow \quad \frac{1-z^5}{(1-z)^5} \equiv 1 \pmod 5,$$

which in particular shows us that,

$$\frac{(1-z^5)(1-z^{10})(1-z^{15})\cdots}{\left\{(1-z)(1-z^2)(1-z^3)\cdots\right\}^5} \equiv 1 \pmod 5.$$

Hence, we easily see that 5 divides each coefficient of z^{5m+1} in the infinite product expansions of,

$$z \cdot \frac{(1-z^5)(1-z^{10})\cdots}{(1-z)(1-z^2)\cdots} = z \cdot \left\{(1-z)(1-z^2)\cdots\right\}^4 \times \frac{(1-z^5)(1-z^{10})\cdots}{\left\{(1-z)(1-z^2)\cdots\right\}^5}.$$

Finally, since we may write the generating function for the partition function as,

$$\begin{aligned}&\frac{z}{(1-z)(1-z^2)\cdots}\\ &= z \cdot \frac{(1-z^5)(1-z^{10})\cdots}{(1-z)(1-z^2)\cdots} \times (1+z^5+z^{10}+\cdots)(1+z^{10}+z^{20}+\cdots)\cdots\\ &= z+\sum_{n\geq 2} p(n-1)z^n,\end{aligned}$$

we may equate the coefficients of z^{5m+5} in the previous equations to prove our desired congruence result, namely that, $p(5m+4)\equiv 0 \pmod 5$ for all $m\geq 0$.

Transformations of Generating Functions

There are a number of transformations of generating functions that provide other applications. A transformation of a sequence's ordinary generating function (OGF) provides a method of converting the generating function for one sequence into a generating function enumerating another. These transformations typically involve integral formulas involving a sequence OGF or weighted sums over the higher-order derivatives of these functions.

Generating function transformations can come into play when we seek to express a generating function for the sums,

$$s_n := \sum_{m=0}^{n} \binom{n}{m} c_{n,m} a_m,$$

in the form of vinvolving the original sequence generating function. For example, if the sums $s_n := \sum_{k\geq 0} \binom{n+k}{m+2k} a_k$, then the generating function for the modified sum expressions is given by $S(z) = \frac{z^m}{(1-z)^{m+1}} A\left(\frac{z}{(1-z)^2}\right)$.

There are also integral formulas for converting between a sequence's OGF, $F(z)$, and its exponential generating function, or EGF, $\hat{F}(z)$, and vice versa given by:

$$F(z) = \int_0^\infty \hat{F}(tz)e^{-t}\,dt$$

$$\hat{F}(z) = \frac{1}{2\eth} \int_{-\eth}^{\eth} F(ze^{-i\vartheta}) e^{e^{i\vartheta}}\, d\vartheta,$$

provided that these integrals converge for appropriate values of z.

Other Applications

Generating functions are used to:

- Find a closed formula for a sequence given in a recurrence relation. For example, consider Fibonacci numbers.
- Find recurrence relations for sequences—the form of a generating function may suggest a recurrence formula.
- Find relationships between sequences—if the generating functions of two sequences have a similar form, then the sequences themselves may be related.
- Explore the asymptotic behaviour of sequences.
- Prove identities involving sequences.
- Solve enumeration problems in combinatorics and encoding their solutions. Rook polynomials are an example of an application in combinatorics.
- Evaluate infinite sums.

Other Generating Functions

Examples:

Examples of polynomial sequences generated by more complex generating functions include:

- Appell polynomials.
- Chebyshev polynomials.
- Difference polynomials.
- Generalized Appell polynomials.
- Q-difference polynomials.

Other sequences generated by more complex generating functions:

- Double exponential generating functions. For example: Aitken's Array: Triangle of Numbers.
- Hadamard products of generating functions/diagonal generating functions and their corresponding integral transformations.

Convolution Polynomials

Convolution Polynomials defines a generalized class of convolution polynomial sequences by their special generating functions of the form,

$$F(z)^x = \exp(x \log F(z)) = \sum_{n \geq 0} f_n(x) z^n,$$

for some analytic function F with a power series expansion such that $F(0)=1$. We say that a family of polynomials, $f_0, f_1, f_2, \ldots$, forms a convolution family if $\deg\{f_n\} \leq n$ and if the following convolution condition holds for all x, y and for all $n \geq 0$:

$$f_n(x+y) = f_n(x)f_0(y) + f_{n-1}(x)f_1(y) + \cdots + f_1(x)f_{n-1}(y) + f_0(x)f_n(y).$$

We see that for non-identically zero convolution families, this definition is equivalent to requiring that the sequence have an ordinary generating function of the first form given above.

A sequence of convolution polynomials defined in the notation above has the following properties:

- The sequence $n! \cdot f_n(x)$ is of binomial type.
- Special values of the sequence include $f_n(1) = [z^n]F(z)$ and $f_n(0) = \delta_{n,0}$.
- For arbitrary (fixed) $x, y, t \in \mathbb{C}$, these polynomials satisfy convolution formulas of the form:

$$f_n(x+y) = \sum_{k=0}^{n} f_k(x) f_{n-k}(y)$$

$$f_n(2x) = \sum_{k=0}^{n} f_k(x) f_{n-k}(x)$$

$$xnf_n(x+y) = (x+y)\sum_{k=0}^{n} kf_k(x) f_{n-k}(y)$$

$$\frac{(x+y)f_n(x+y+tn)}{x+y+tn} = \sum_{k=0}^{n} \frac{xf_k(x+tk)}{x+tk} \frac{yf_{n-k}(y+t(n-k))}{y+t(n-k)}.$$

For a fixed non-zero parameter $t \in \mathbb{C}$, we have modified generating functions for these convolution polynomial sequences given by,

$$\frac{zF_n(x+tn)}{(x+tn)} = [z^n]\mathcal{F}_t(z)^x,$$

where $\mathcal{F}_t(z)$ is implicitly defined by a functional equation of the form $\mathcal{F}_t(z) = F(x\mathcal{F}_t(z)^t)$. Moreover, we can use matrix methods (as in the reference) to prove that given two convolution polynomial sequences, $\langle f_n(x) \rangle$ and $\langle g_n(x) \rangle$, with respective corresponding generating functions, $F(z)^x$ and $G(z)^x$, then for arbitrary t we have the identity,

$$[z^n]\left(G(z)F\left(zG(z)^t\right)\right)^x = \sum_{k=0}^{n} F_k(x)G_{n-k}(x+tk).$$

Examples of convolution polynomial sequences include the binomial power series, $\mathcal{B}_t(z) = 1 + z\mathcal{B}_t(z)^t$, so-termed tree polynomials, the Bell numbers, $B(n)$, the Laguerre polynomials, and the Stirling convolution polynomials.

Tables of Special Generating Functions

An initial listing of special mathematical series is found here. Other special generating functions of note include the entries in the next table, which is by no means complete.

Formal power series	Generating-function formula	Notes
$\sum_{n\geq 0}\binom{m+n}{n}(H_{n+m}-H_m)z^n$	$\frac{1}{(1-z)^{m+1}}\ln\frac{1}{1-z}$	H_n is a first-order harmonic number.
$\sum_{n\geq 0} B_n \frac{z^n}{n!}$	$\frac{z}{e^z-1}$	B_n is a Bernoulli number.
$\sum_{n\geq 0} F_{mn}z^n$	$\frac{F_m z}{1-(F_{m-1}+F_{m+1})z+(-1)^m z^2}$	F_n is a Fibonacci number and $m \in \mathbb{Z}^+$
$\sum_{n\geq 0}\left\{ {n \atop m} \right\} z^n$	$(z^{-1})^{\overline{-m}} = \frac{z^m}{(1-z)(1-2z)\cdots(1-mz)}$	$x^{\overline{n}}$ denotes the rising factorial, or Pochhammer symbol and some integer $m \geq 0$
$\sum_{n\geq 0}\left[{n \atop m} \right] z^n$	$z^{\overline{m}} = z(z+1)\cdots(z+m-1)$	
$\sum_{n\geq 1}\frac{(-1)^{n-1}4^n(4^n-2)B_{2n}z^{2n}}{(2n)\cdot(2n)!}$	$\ln\frac{\tan(z)}{z}$	
$\sum_{n\geq 0}\frac{(1/2)^{\overline{n}}z^{2n}}{(2n+1)\cdot n!}$	$z^{-1}\arcsin(z)$	
$\sum_{n\geq 0} H_n^{(s)}z^n$	$\frac{\mathrm{Li}_s(z)}{1-z}$	$\mathrm{Li}_s(z)$ is the polylogarithm function and $H_n^{(s)}$ is a generalized harmonic number for $\Re(s) > 1$
$\sum_{n\geq 0} n^m z^n$	$\sum_{0\leq j\leq m}\left\{ {m \atop j} \right\}\frac{j!\cdot z^j}{(1-z)^{j+1}}$	$\left\{ {n \atop m} \right\}$ is a Stirling number of the second kind and where the individual terms in the expansion satisfy $\frac{z^i}{(1-z)^{i+1}} = \sum_{k=0}^{i}\binom{i}{k}\frac{(-1)^{k-i}}{(1-z)^{k+1}}$
$\sum_{k<n}\binom{n-k}{k}\frac{n}{n-k}z^k$	$\left(\frac{1+\sqrt{1+4z}}{2}\right)^n + \left(\frac{1-\sqrt{1+4z}}{2}\right)^n$	

$\sum_{n_1,\ldots,n_m\geq 0} \min(n_1,\ldots,n_m) z_1^{n_1}\cdots z_m^{n_m}$	$\frac{z_1\cdots z_m}{(1-z_1)\cdots(1-z_m)(1-z_1\cdots z_m)}$	The two-variable case is given by: $M(w,z) := \sum_{m,n\geq 0} \min(m,n) w^m z^n = \frac{wz}{(1-w)(1-z)(1-wz)}$
$\sum_{n\geq 0} \binom{s}{n} z^n$	$(1+z)^s$	$s \in \mathbb{C}$
$\sum_{n\geq 0} \binom{n}{k} z^n$	$\frac{z^k}{(1-z)^{k+1}}$	$k \in \mathbb{N}$
$\sum_{n\geq 1} \log(n) z^n$	$-\frac{\partial}{\partial s} \mathrm{Li}_s(z)\|_{s=0}$	

GENERATING FUNCTION TRANSFORMATION

In mathematics, a transformation of a sequence's generating function provides a method of converting the generating function for one sequence into a generating function enumerating another. These transformations typically involve integral formulas applied to a sequence generating function or weighted sums over the higher-order derivatives of these functions.

Given a sequence, $\{f_n\}_{n=0}^{\infty}$, the ordinary generating function (OGF) of the sequence, denoted $F(z)$, and the exponential generating function (EGF) of the sequence, denoted $\hat{F}(z)$, are defined by the formal power series,

$$F(z) = \sum_{n=0}^{\infty} f_n z^n = f_0 + f_1 z + f_2 z^2 + \cdots$$

$$\hat{F}(z) = \sum_{n=0}^{\infty} \frac{f_n}{n!} z^n = \frac{f_0}{0!} + \frac{f_1}{1!} z + \frac{f_2}{2!} z^2 + \cdots.$$

We use the convention that the ordinary (exponential) generating function for a sequence $\{f_n\}$ is denoted by the uppercase function $F(z)/\hat{F}(z)$ for some fixed or formal z when the context of this notation is clear. Additionally, we use the bracket notation for coefficient extraction which is given by $[z^n]F(z) := f_n$.

Extracting Arithmetic Progressions of a Sequence

The focus is to give formulas for generating functions enumerating the sequence $\{f_{an+b}\}$ given an ordinary generating function $F(z)$ where $a, b \in \mathbb{N}$, $a \geq 2$, and $\leq b < a$. In the first two cases where $(a,b) := (2,0),(2,1)$, we can expand these arithmetic progression generating functions directly in terms of $F(z)$:

$$\sum_{n\geq 0} f_{2n} z^{2n} = \frac{1}{2}(F(z)+F(-z))$$

$$\sum_{n\geq 0} f_{2n+1} z^{2n+1} = \frac{1}{2}(F(z)-F(-z)).$$

More generally, suppose that $a \geq 3$ and that $\omega_a \equiv \exp\left(\frac{2\pi\iota}{a}\right)$ denotes the ath primitive root of unity. Then we have the formula:

$$a^{th} \sum_{n\geq 0} f_{an+b} z^{an+b} = \frac{1}{a} \times \sum_{m=0}^{a-1} \omega_a^{-mb} F\left(\omega_a^m z\right).$$

For integers $m \geq 1$, another useful formula providing somewhat reversed floored arithmetic progressions are generated by the identity,

$$\sum_{n\geq 0} f_{\left\lfloor \frac{n}{m} \right\rfloor} z^n = \frac{1-z^m}{1-z} F(z^m) = (1+z+\cdots+z^{m-2}+z^{m-1})F(z^m).$$

Powers of an OGF and Composition with Functions

The exponential Bell polynomials, $B_{n,k}(x_1,\ldots,x_n) := n! \cdot [t^n u^k]\Phi(t,u)$, are defined by the exponential generating function,

$$\Phi(t,u) = \exp\left(u \times \sum_{m\geq 1} x_m \frac{t^m}{m!}\right) = 1 + \sum_{n\geq 1} \{\sum_{k=1}^{n} B_{n,k}(x_1,x_2,\ldots)u^k\} \frac{t^n}{n!}.$$

The next formulas for powers, logarithms, and compositions of formal power series are expanded by these polynomials with variables in the coefficients of the original generating functions. The formula for the exponential of a generating function is given implicitly through the Bell polynomials by the EGF for these polynomials defined in the previous formula for some sequence of $\{x_i\}$.

Reciprocals of an OGF (Special Case of the Powers Formula)

The power series for the reciprocal of a generating function, F(z), is expanded by,

$$\frac{1}{F(z)} = \frac{1}{f_0} - \frac{f_1}{f_0^2} z + \frac{\left(f_1^2 - f_0 f_2\right)}{f_0^3} z^2 - \frac{f_1^3 - 2f_0 f_1 f_2 + f_0^2 f_3}{f_0^4} + \cdots.$$

If we let $b_n := [z^n]1/F(z)$ denote the coefficients in the expansion of the reciprocal generating function, then we have the following recurrence relation:

$$b_n = -\frac{1}{f_0}(f_1 b_{n-1} + f_2 b_{n-2} + \cdots + f_n b_0), n \geq 1.$$

Powers of an OGF

Let $m \in \mathbb{C}$ be fixed, suppose that $f_0 = 1$, and denote $b_n^{(m)} := [z^n]F(z)^m$. Then we have a series expansion for $F(z)^m$ given by:

$$F(z)^m = 1 + mf_1 z + m\left((m-1)f_1^2 + 2f_2\right)\frac{z^2}{2} + \left(m(m-1)(m-2)f_1^3 + 6m(m-1)f_2 + 6mf_3\right)\frac{z^3}{6} + \cdots,$$

and the coefficients $b_n^{(m)}$ satisfy a recurrence relation of the form,

$$n \cdot b_n^{(m)} = (m-n+1)f_1 b_{n-1}^{(m)} + (2m-n+2)f_2 b_{n-2}^{(m)} + \cdots + ((n-1)m-1)f_{n-1} b_1^{(m)} + nmf_n, n \geq 1.$$

Another formula for the coefficients, $b_n^{(m)}$, is expanded by the Bell polynomials as,

$$F(z)^m = f_0^m + \sum_{n \geq 1}\left(\sum_{1 \leq k \leq n} (m)_k f_0^{m-k} B_{n,k}(f_1 \cdot 1!, f_2 \cdot 2!, \ldots)\right)\frac{z^n}{n!},$$

where $(r)_n$ denotes the Pochhammer symbol.

Logarithms of an OGF

If we let $f_0 = 1$ and define $q_n := [z^n]\log F(z),$, then we have a power series expansion for the composite generating function given by:

$$\log F(z) = f_1 + \left(2f_2 - f_1^2\right)\frac{z}{2} + \left(3f_3 - 3f_1 f_2 + f_1^3\right)\frac{z^2}{3} + \cdots,$$

where the coefficients, q_n, in the previous expansion satisfy the recurrence relation given by,

$$n \cdot q_n = nf_n - (n-1)f_1 q_{n-1} - (n-2)f_2 q_{n-2} - \cdots - f_{n-1} q_1,$$

and a corresponding formula expanded by the Bell polynomials in the form of the power series coefficients of the following generating function:

$$\log F(z) = \sum_{n \geq 1}\left(\sum_{1 \leq k \leq n} (-1)^{k-1}(k-1)! B_{n,k}(f_1 \cdot 1!, f_2 \cdot 2!, \ldots)\right)\frac{z^n}{n!}.$$

Faà di Bruno's Formula

Let $\hat{F}(z)$ denote the EGF of the sequence, $\{f_n\}_{n\geq 0}$, and suppose that $\hat{G}(z)$ is the EGF of the sequence, $\{g_n\}_{n\geq 0}$. The sequence, $\{h_n\}_{n\geq 0}$, generated by the exponential generating function for the composition, $\hat{H}(z) := \hat{F}(\hat{G}(z))$, is given in terms of the exponential Bell polynomials as follows:

$$h_n = \sum_{1 \leq k \leq n} f_k \cdot B_{n,k}(g_1, g_2, \cdots, g_{n-k+1}) + f_0 \cdot \delta_{n,0}.$$

We compare the statement of this result to the other known statement of Faà di Bruno's formula which provides an analogous expansion of the nth derivatives of a composite function in terms of the derivatives of the two functions of *z* defined.

Integral Transformations

OGF ↔ EGF Conversion Formulas

We have the following integral formulas for $a, b \in \mathbb{Z}^+$ which can be applied termwise with respect to z when z is taken to be any formal power series variable:

$$\sum_{n\geq 0} f_n z^n = \int_0^\infty \hat{F}(tz)e^{-t}dt$$

$$\sum_{n\geq 0} \Gamma(an+b)\cdot f_n z^n = \int_0^\infty t^{b-1}e^{-t}F(t^a z)dt.$$

$$\sum_{n\geq 0} \frac{f_n}{n!} z^n = \frac{1}{2\pi}\int_{-\pi}^{\pi} F\left(ze^{-\imath\vartheta}\right)e^{e^{\imath\vartheta}}d\vartheta.$$

Notice that the first and last of these integral formulas are used to convert between the EGF to the OGF of a sequence, and from the OGF to the EGF of a sequence whenever these integrals are convergent.

The first integral formula corresponds to the Laplace transform (or sometimes the formal Laplace–Borel transformation) of generating functions, denoted by $\mathcal{L}[F](z)$, defined in. Other integral representations for the gamma function in the second of the previous formulas can of course also be used to construct similar integral transformations. One particular formula results in the case of the double factorial function example given immediately below. The last integral formula is compared to Hankel's loop integral for the reciprocal gamma function applied termwise to the power series for $F(z)$.

Example: A Double Factorial Integral for the EGF of the Stirling Numbers of the Second Kind

The single factorial function, $(2n)!$, is expressed as a product of two double factorial functions of the form:

$$(2n)! = (2n)!! \times (2n-1)!! = \frac{4^n \cdot n!}{\sqrt{\pi}} \times \Gamma\left(n + \frac{1}{2}\right),$$

where an integral for the double factorial function, or rational gamma function, is given by,

$$\frac{1}{2}\cdot(2n-1)!! = \frac{2^n}{\sqrt{4\pi}}\Gamma\left(n+\frac{1}{2}\right) = \frac{1}{\sqrt{2\pi}} \times \int_0^\infty e^{-t^2/2}t^{2n}\,dt,$$

for natural numbers $n \geq 0$. This integral representation of $(2n-1)!!$ then implies that for fixed non-zero $q \in \mathbb{C}$ and any integral powers $k \geq 0$, we have the formula,

$$\frac{\log(q)^k}{k!} = \frac{1}{(2k)!} \times \left[\int_0^\infty \frac{2e^{-t^2/2}}{\sqrt{2\pi}} (\sqrt{2\log(q)} \cdot t)^{2k} \, dt \right].$$

Thus for any prescribed integer $j \geq 0$, we can use the previous integral representation together with the formula for extracting arithmetic progressions from a sequence OGF given above, to formulate the next integral representation for the so-termed modified Stirling number EGF as,

$$\sum_{n \geq 0} \left\{ \begin{matrix} 2n \\ j \end{matrix} \right\} \frac{\log(q)^n}{n!} = \int_0^\infty \frac{e^{-t^2/2}}{\sqrt{2\pi} \cdot j!} \left[\sum_{b = \pm 1} \left(e^{b\sqrt{2\log(q)} \cdot t} - 1 \right)^j \right] dt,$$

which is convergent provided suitable conditions on the parameter $0 < |q| < 1$.

Example: An EGF Formula for the Higher-order Derivatives of the Geometric Series

For fixed non-zero $c, z \in \mathbb{C}$ defined such that $|cz| < 1$, let the geometric series over the non-negative integral powers of $(cz)^n$ be denoted by $G(z) := 1/(1-cz)$. The corresponding higher-order j^{th} derivatives of the geometric series with respect to z are denoted by the sequence of functions,

$$G_j(z) := \frac{(cz)^j}{1-cz} \times \left(\frac{d}{dz} \right)^{(j)} [G(z)],$$

for non-negative integers $j \geq 0$. These j^{th} derivatives of the ordinary geometric series can be shown, for example by induction, to satisfy an explicit closed-form formula given by,

$$G_j(z) = \frac{(cz)^j \cdot j!}{(1-cz)^{j+2}},$$

for any $j \geq 0$. whenever $|cz| < 1$. As an example of the third OGF $\mapsto$ EGF conversion formula cited above, we can compute the following corresponding exponential forms of the generating functions $G_j(z)$:

$$\hat{G}_j(z) = \frac{1}{2\pi} \int_{-\pi}^{+\pi} G_j\left(ze^{-\imath t}\right) e^{e^{\imath t}} \, dt = \frac{(cz)^j e^{cz}}{(j+1)} (j+1+z).$$

Fractional Integrals and Derivatives

Fractional integrals and fractional derivatives form another generalized class of integration and differentiation operations that can be applied to the OGF of a sequence to form the corresponding OGF of a transformed sequence. For $\Re(\alpha) > 0$ we define the fractional integral operator (of order α) by the integral transformation,

$$I^{\alpha}F(z) = \frac{1}{\Gamma(\alpha)}\int_0^z (z-t)^{\alpha-1}F(t)dt,$$

which corresponds to the (formal) power series given by,

$$I^{\alpha}F(z) = \sum_{n\geq 0}\frac{n!}{\Gamma(n+\alpha+1)}f_n z^{n+\alpha}.$$

For fixed $\alpha\ \beta \in \mathbb{C}$ defined such that $\Re(\alpha), \Re(\beta) > 0$, we have that the operators $I^{\alpha}I^{\beta} = I^{\alpha+\beta}$. Moreover, for fixed $\alpha \in \mathbb{C}$ and integers n satisfying $0 < \Re(\alpha) < n$ we can define the notion of the fractional derivative satisfying the properties that,

$$D^{\alpha}F(z) = \frac{d^{(n)}}{dz^{(n)}} I^{n-\alpha}F(z),$$

and $D^k I^{\alpha} = D^n I^{\alpha+n-k}$ for $k = 1,2,\ldots,n$,

where we have the semigroup property that $D^{\alpha}D^{\beta} = D^{\alpha+\beta}$ only when none of $\alpha, \beta, \alpha+\beta$ is integer-valued.

Polylogarithm Series Transformations

For fixed $s \in \mathbb{Z}^+$, we have that (compare to the special case of the integral formula for the Nielsen generalized polylogarithm function defined in).

$$\sum_{n\geq 0}\frac{f_n}{(n+1)^s}z^n = \frac{(-1)^{s-1}}{(s-1)!}\int_0^1 \log^{s-1}(t)F(tz)dt.$$

Notice that if we set $g_n \equiv f_{n+1}$, the integral with respect to the generating function, $G(z)$, in the last equation when $z \equiv 1$ corresponds to the Dirichlet generating function, or DGF, $\tilde{F}(s)$,, of the sequence of $\{f_n\}$ provided that the integral converges. This class of polylogarithm-related integral transformations is related to the derivative-based zeta series transformations.

Square Series Generating Function Transformations

For fixed non-zero $q, c, z \in \mathbb{C}$ such that $|q| < 1$ and $|cz| < 1$, we have the following integral representations for the so-termed square series generating function associated with the sequence $\{f_n\}$, which can be integrated termwise with respect to z:

$$\sum_{n\geq 0} q^{n^2} f_n \cdot (cz)^n = \frac{1}{\sqrt{2\pi}}\int_0^{\infty} e^{-t^2/2}\left[F\left(e^{t\sqrt{2\log(q)}}\cdot cz\right) + F\left(e^{-t\sqrt{2\log(q)}}\cdot cz\right)\right]dt.$$

This result, which is proved in the reference, follows from a variant of the double factorial function transformation integral for the Stirling numbers of the second kind given as an example above. In particular, since,

$$q^{n^2} = \exp(n^2 \cdot \log(q)) = 1 + n^2 \log(q) + n^4 \frac{\log(q)^2}{2!} + n^6 \frac{\log(q)^3}{3!} + \cdots,$$

we can use a variant of the positive-order derivative-based OGF transformations defined in the next sections involving the Stirling numbers of the second kind to obtain an integral formula for the generating function of the sequence, $\{S(2n,j)/n!\}$, and then perform a sum over the jth derivatives of the formal OGF, $F(z)$ to obtain the result in the previous equation where the arithmetic progression generating function at hand is denoted by:

$$\sum_{n\geq 0} \left\{ \begin{matrix} 2n \\ j \end{matrix} \right\} \frac{z^{2n}}{(2n)!} = \frac{1}{2j!}((e^z - 1)^j + (e^{-z} - 1)^j),$$

for each fixed $j \in \mathbb{N}$.

Hadamard Products and Diagonal Generating Functions

We have an integral representation for the Hadamard product of two generating functions, $F(z)$ and $G(z)$, stated in the following form:

$$(F \odot G)(z) := \sum_{n\geq 0} f_n g_n z^n = \frac{1}{2\pi} \int_0^{2\pi} F\left(\sqrt{z} e^{it}\right) G\left(\sqrt{z} e^{-it}\right) dt.$$

The reference also provides nested coefficient extraction formulas of the form:

$$\operatorname{diag}(F_1 \cdots F_k) := \sum_{n\geq 0} f_{1,n} \cdots f_{k,n} z^n = [x_{k-1}^0 \cdots x_2^0 x_1^0] F_k\left(\frac{z}{x_{k-1}}\right) F_{k-1}\left(\frac{x_{k-1}}{x_{k-2}}\right) \cdots F_2\left(\frac{x_2}{x_1}\right) F_1(x_1),$$

which are particularly useful in the cases where the component sequence generating functions, $F_i(z)$, can be expanded in a Laurent series, or fractional series, in z, such as in the special case where all of the component generating functions are rational, which leads to an algebraic form of the corresponding diagonal generating function.

Example: Hadamard Products of Rational Generating Functions

In general, the Hadamard product of two rational generating functions is itself rational. This is seen by noticing that the coefficients of a rational generating function form quasi-polynomial terms of the form:

$$f_n = p_1(n)\rho_1^n + \cdots + p_\ell(n)\rho_\ell^n,$$

where the reciprocal roots, $\rho_i \in \mathbb{C}$, are fixed scalars and where $p_i(n)$ is a polynomial in for all $1 \leq i \leq \ell$. For example, the Hadamard product of the two generating functions,

$$F(z) := \frac{1}{1 + a_1 z + a_2 z^2}$$

and, $G(z) := \dfrac{1}{1 + b_1 z + b_2 z^2}$
is given by the rational generating function formula,

$$(F \odot G)(z) = \frac{1 - a_2 b_2 z^2}{1 - a_1 b_1 z + \left(a_2 b_1^2 + a_1^2 b_2 - a_2 b_2\right) z^2 - a_1 a_2 b_1 b_2 z^3 + a_2^2 b_2^2 z^4}.$$

Example: Factorial (Approximate Laplace) Transformations

Ordinary generating functions for generalized factorial functions formed as special cases of the generalized rising factorial product functions, or Pochhammer k-symbol, defined by,

$$p_n(\alpha, R) := R(R+\alpha)\cdots(R+(n-1)\alpha) = \alpha^n \cdot \left(\frac{R}{\alpha}\right)_n,$$

where R is fixed, $\alpha \neq 0$, and $(x)_n$ denotes the Pochhammer symbol are generated (at least formally) by the Jacobi-type J-fractions (or special forms of continued fractions) established in the reference. If we let $\mathrm{Conv}_h(\alpha, R; z) := \mathrm{FP}_h(\alpha, R; z) / \mathrm{FQ}_h(\alpha, R; z)$ denote the hth convergent to these infinite continued fractions where the component convergent functions are defined for all integers $h \geq 2$ by,

$$\mathrm{FP}_h(\alpha, R; z) = \sum_{n=0}^{h-1} \left[\sum_{k=0}^{n} \binom{h}{k} \left(1 - h - \frac{R}{\alpha}\right)_k \left(\frac{R}{\alpha}\right)_{n-k} \right] (\alpha z)^n,$$

and,

$$\begin{aligned} \mathrm{FQ}_h(\alpha, R; z) &= (-\alpha z)^h \cdot h! \times L_h^{(R/\alpha - 1)}\left((\alpha z)^{-1}\right) \\ &= \sum_{k=0}^{h} \binom{h}{k} \left[\prod_{j=0}^{k-1} (R + (j - 1 - j)\alpha) \right] (-z)^k, \end{aligned}$$

where $L_n^{(\beta)}(x)$ denotes an associated Laguerre polynomial, then we have that the hth convergent function, $\mathrm{Conv}_h(\alpha, R; z)$, exactly enumerates the product sequences, $p_n(\alpha, R)$, for all $0 \leq n < 2h$. For each $h \geq 2$, the hth convergent function is expanded as a finite sum involving only paired reciprocals of the Laguerre polynomials in the form of,

$$\mathrm{Conv}_h(\alpha, R; z) = \sum_{i=0}^{h-1} \binom{\frac{R}{\alpha} + i - 1}{i} \times \frac{(-\alpha z)^{-1}}{(i+1) \cdot L_i^{(R/\alpha - 1)}\left((\alpha z)^{-1}\right) L_{i+1}^{(R/\alpha - 1)}\left((\alpha z)^{-1}\right)}$$

Moreover, since the single factorial function is given by both $n! = p_n(1,1)$ and $n! = p_n(-1,n)$, we can generate the single factorial function terms using the approximate rational convergent generating functions up to order $2h$. This observation suggests an approach to approximating the exact (formal) Laplace–Borel transform usually given in terms of the integral representation by a Hadamard product, or diagonal-coefficient, generating function. In particular, given any OGF $G(z)$ we can form the approximate Laplace transform, which is $2h$-order accurate, by the diagonal coefficient extraction formula stated above given by:

$$\begin{aligned}\tilde{\mathcal{L}}_h[G](z) &:= [x^0]\operatorname{Conv}_h\left(1,1;\frac{z}{x}\right)G(x) \\ &= \frac{1}{2\pi}\int_0^{2\pi}\operatorname{Conv}_h\left(1,1;\sqrt{z}e^{\imath t}\right)G\left(-\sqrt{z}e^{\imath t}\right)dt.\end{aligned}$$

Examples of sequences enumerated through these diagonal coefficient generating functions arising from the sequence factorial function multiplier provided by the rational convergent functions include,

$$n!^2 = \left[z^n\right]\left[x^0\right]\operatorname{Conv}_h\left(-1,n;\frac{z}{x}\right)\operatorname{Conv}_h(-1,n;x), h \geq n$$

$$\binom{2n}{n} = \left[x_1^0 x_2^0 z^n\right]\operatorname{Conv}_h\left(-2,2n;\frac{z}{x_2}\right)\operatorname{Conv}_h\left(-2,2n-1;\frac{x_2}{x_2}\right)I_0\left(2\sqrt{x_1}\right)$$

$$\binom{3n}{n}\binom{2n}{n} = \left[x_1^0 x_2^0 z^n\right]\operatorname{Conv}_h\left(-3,3n-1;\frac{3_z}{x_2}\right)\operatorname{Conv}_h\left(-3,3n-2;\frac{x_2}{x_1}\right)I_0\left(2\sqrt{x_1}\right)$$

$$!n = n! \times \sum_{i=0}^{n}\frac{(-1)^i}{i!} = \left[z^n x^0\right]\left(\frac{e^{-x}}{(1-x)}\operatorname{Conv}_n\left(-1,n;\frac{z}{x}\right)\right)$$

$$\operatorname{af}(n) = \sum_{k=1}^{n}(-1)^{n-k}k! = \left[z^n\right]\left(\frac{\operatorname{Conv}_n(1,1:z)-1}{1+z}\right)$$

$$\begin{aligned}(t-1)^n P_n\left(\frac{t+1}{t-1}\right) &= \sum_{k=0}^{n}\binom{n}{k}^2 t^k \\ &= \left[x_1^0 x_2^0\right]\left[z^n\right]\left(\operatorname{Conv}_n\left(1,1;\frac{z}{x_1}\right)\operatorname{Conv}_n\left(1,1;\frac{x_1}{x_2}\right)I_0\left(2\sqrt{t.x_2}\right)I_0\left(2\sqrt{x_2}\right)\right), n \geq 1\end{aligned}$$

$$\begin{aligned}(2n-1)!! &= \sum_{k=1}^{n}\frac{(n-1)!}{(k-1)!}k\cdot(2k-3)!! \\ &= \left[x_1^0 x_2^0 x_3^{n-1}\right]\left(\operatorname{Conv}_n\left(1,1;\frac{x_3}{x_2}\right)\operatorname{Conv}_n\left(2,1;\frac{x_2}{x_1}\right)\frac{(x_1+1)e^{x_1}}{1-x_2}\right),\end{aligned}$$

where $I_0(z)$ denotes a modified Bessel function, !n denotes the subfactorial function, af(n) denotes the alternating factorial function, and $P_n(x)$ is a Legendre polynomial.

Other examples of sequences enumerated through applications of these rational Hadamard product generating functions include the Barnes G-function, combinatorial sums involving the double factorial function, sums of powers sequences, and sequences of binomials.

Derivative Transformations

Positive and Negative-order Zeta Series Transformations

For fixed $k \in \mathbb{Z}^+$, we have that if the sequence OGF $F(z)$ has j^{th} derivatives of all required orders for $1 \le j \le k$, that the positive-order zeta series transformation is given by:

$$\sum_{n \ge 0} n^k f_n z^n = \sum_{j=0}^{k} \left\{ {k \atop j} \right\} z^j F^{(j)}(z),$$

where $\left\{ {n \atop k} \right\}$ denotes a Stirling number of the second kind. In particular, we have the following special case identity when $f_n \equiv 1 \forall n$ when $\left\langle {n \atop m} \right\rangle$ denotes the triangle of first-order Eulerian numbers:

$$\sum_{n \ge 0} n^k z^n = \sum_{j=0}^{k} \left\{ {k \atop j} \right\} \frac{z^j \cdot j!}{(1-z)^{j+1}} = \frac{1}{(1-z)^{k+1}} \times \sum_{0 \le m < k} \left\langle {k \atop m} \right\rangle z^{m+1}.$$

We can also expand the negative-order zeta series transformations by a similar procedure to the above expansions given in terms of the jth-order derivatives of some $F(z) \in C^{\infty}$ and an infinite, non-triangular set of generalized Stirling numbers in reverse, or generalized Stirling numbers of the second kind defined within this context.

In particular, for integers $k, j \ge 0$, define these generalized classes of Stirling numbers of the second kind by the formula,

$$\left\{ {k+2 \atop j} \right\}_* := \frac{1}{j!} \times \sum_{m=1}^{j} \binom{j}{m} \frac{(-1)^{j-m}}{m^k}.$$

Then for $k \in \mathbb{Z}^+$ and some prescribed OGF, $F(z) \in C^{\infty}$, i.e., so that the higher-order j^{th} derivatives of $F(z)$ exist for all $j \ge 0$, we have that,

$$\sum_{n \ge 1} \frac{f_n}{n^k} z^n = \sum_{j \ge 1} \left\{ {k+2 \atop j} \right\}_* z^j F^{(j)}(z).$$

A table of the first few zeta series transformation coefficients, $\left\{ {k \atop j} \right\}_*$, appears below.

These weighted-harmonic-number expansions are almost identical to the known formulas for the Stirling numbers of the first kind up to the leading sign on the weighted harmonic number terms in the expansions.

K	$\left\{ \begin{matrix} k \\ j \end{matrix} \right\}_* \times (-1)^{j-1} j!$
2	1
3	H_J
4	$\frac{1}{2}\left(H_j^2 + H_j^{(2)}\right)$
5	$\frac{1}{6}\left(H_j^3 + 3H_jH_j^{(2)} + 2H_j^{(3)}\right)$
6	$\frac{1}{24}\left(H_j^4 + 6H_j^2H_j^{(2)} + 3\left(H_j^{(2)}\right)^2 + 8H_jH_j^{(3)} + 6H_j^{(4)}\right)$

Examples of the Negative-order Zeta Series Transformations

The next series related to the polylogarithm functions (the dilogarithm and trilogarithm functions, respectively), the alternating zeta function and the Riemann zeta function are formulated from the previous negative-order series results found in the references. In particular, when $s := 2$ (or equivalently, when $k := 4$ in the table above), we have the following special case series for the dilogarithm and corresponding constant value of the alternating zeta function:

$$\mathrm{Li}_2(z) = \sum_{j\geq 1} \frac{(-1)^{j-1}}{2}\left(H_j^2 + H_j^{(2)}\right)\frac{z^j}{(1-z)^{j+1}}$$

$$\zeta^*(2) = \frac{\pi^2}{12} = \sum_{j\geq 1} \frac{\left(H_j^2 + H_j^{(2)}\right)}{4\cdot 2^j}.$$

When $s := 3$ (or when $k := 5$ in the notation, we similarly obtain special case series for these functions given by,

$$\mathrm{Li}_3(z) = \sum_{j\geq 1} \frac{(-1)^{j-1}}{6}\left(H_j^3 + 3H_jH_j^{(2)} + 2H_j^{(3)}\right)\frac{z^j}{(1-z)^{j+1}}$$

$$\zeta^*(3) = \frac{3}{4}\zeta(3) = \sum_{j\geq 1} \frac{\left(H_j^3 + 3H_jH_j^{(2)} + 2H_j^{(3)}\right)}{12\cdot 2^j}$$

$$= \frac{1}{6}\log(2)^3 + \sum_{j\geq 0} \frac{H_jH_j^{(2)}}{2^{j+1}}.$$

It is known that the first-order harmonic numbers have a closed-form exponential generating function expanded in terms of the natural logarithm, the incomplete gamma function, and the exponential integral given by:

$$\sum_{n\geq 0} \frac{H_n}{n!} z^n = e^z\left(E_1(z) + \gamma + \log z\right) = e^z\left(\Gamma(0,z) + \gamma + \log z\right).$$

Additional series representations for the r-order harmonic number exponential generating functions for integers $r \geq 2$ are formed as special cases of these negative-order derivative-based series transformation results. For example, the second-order harmonic numbers have a corresponding exponential generating function expanded by the series,

$$\sum_{n \geq 0} \frac{H_n^{(2)}}{n!} z^n = \sum_{j \geq 1} \frac{H_j^2 + H_j^{(2)}}{2 \cdot (j+1)!} z^j e^z (j+1+z).$$

Generalized Negative-order Zeta Series Transformations

A further generalization of the negative-order series transformations defined above is related to more Hurwitz-zeta-like, or Lerch-transcendent-like, generating functions. Specifically, if we define the even more general parametrized Stirling numbers of the second kind by:

$$\left\{ \begin{matrix} k+2 \\ j \end{matrix} \right\}_{(\alpha,\beta)^*} := \frac{1}{j!} \times \sum_{0 \leq m \leq j} \binom{j}{m} \frac{(-1)^{j-m}}{(\alpha m + \beta)^k},$$

for non-zero $\alpha, \beta \in \mathbb{C}$ such that $-\frac{\beta}{\alpha} \notin \mathbb{Z}^+$, and some fixed $k \geq 1$, we have that,

$$\sum_{n \geq 1} \frac{f_n}{(\alpha n + \beta)^k} z^n = \sum_{j \geq 1} \left\{ \begin{matrix} k+2 \\ j \end{matrix} \right\}_{(\alpha,\beta)^*} z^j F^{(j)}(z).$$

Moreover, for any integers $u, u_0 \geq 0$, we have the partial series approximations to the full infinite series in the previous equation given by:

$$\sum_{n=1}^{u} \frac{f_n}{(\alpha n + \beta)^k} z^n = [w^u] \left(\sum_{j=1}^{u+u_0} \left\{ \begin{matrix} k+2 \\ j \end{matrix} \right\}_{(\alpha,\beta)^*} \frac{(wz)^j F^{(j)}(wz)}{1-w} \right).$$

Examples of the Generalized Negative-order Zeta Series Transformations

Series for special constants and zeta-related functions resulting from these generalized derivative-based series transformations typically involve the generalized r-order harmonic numbers defined by $H_n^{(r)}(\alpha,\beta) := \sum_{1 \leq k \leq n} (\alpha k + \beta)^{-r}$ for integers $r \geq 1$. A pair of particular series expansions for the following constants when $n \in \mathbb{Z}^+$ is fixed follow from special cases of BBP-type identities as:

$$\frac{4\sqrt{3}\pi}{9} = \sum_{j \geq 0} \frac{8}{9^{j+1}} \left(2 \binom{j+\frac{1}{3}}{\frac{1}{3}}^{-1} + \frac{1}{2} \binom{j+\frac{2}{3}}{\frac{2}{3}}^{-1} \right)$$

$$\log\left(\frac{n^2-n+1}{n^2}\right) = \sum_{j \geq 0} \frac{1}{(n^2+1)^{j+1}} \left(\frac{2}{3\cdot(j+1)} - n^2 \binom{j+\frac{1}{3}}{\frac{1}{3}}^{-1} + \frac{n}{2} \binom{j+\frac{2}{3}}{\frac{2}{3}}^{-1} \right).$$

Several other series for the zeta-function-related cases of the Legendre chi function, the polygamma function, and the Riemann zeta function include,

$$\chi_1(z) = \sum_{j \geq 0} \binom{j+\frac{1}{2}}{\frac{1}{2}}^{-1} \frac{z \cdot (-z^2)^j}{(1-z^2)^{j+1}}$$

$$\chi_2(z) = \sum_{j \geq 0} \binom{j+\frac{1}{2}}{\frac{1}{2}}^{-1} \left(1 + H_j^{(1)}(2,1)\right) \frac{z \cdot (-z^2)^j}{(1-z^2)^{j+1}}$$

$$\sum_{k \geq 0} \frac{(-1)^k}{(z+k)^2} = \sum_{j \geq 0} \binom{j+z}{z}^{-1} \left(\frac{1}{z^2} + \frac{1}{z} H_j^{(1)}(2,z) \right) \frac{1}{2^{j+1}}$$

$$\frac{13}{18}\zeta(3) = \sum_{i=1,2} \sum_{j \geq 0} \binom{j+\frac{i}{3}}{\frac{i}{3}}^{-1} \left(\frac{1}{i^3} + \frac{1}{i^2} H_j^{(1)}(3,i) + \frac{1}{2i}\left(H_j^{(1)}(3,i)^2 + H_j^{(2)}(3,i)\right) \right) \frac{(-1)^{i+1}}{2^{j+1}}.$$

Additionally, we can give another new explicit series representation of the inverse tangent function through its relation to the Fibonacci numbers expanded as in the references by,

$$\tan^{-1}(x) = \frac{\sqrt{5}}{2\acute{y}} \times \sum_{b=\pm 1} \sum_{j \geq 0} \frac{b}{\sqrt{5}} \binom{j+\frac{1}{2}}{j}^{-1} \left[\frac{(b\iota\varphi t/\sqrt{5})^j}{\left(1 - \frac{b\iota\varphi t}{\sqrt{5}}\right)^{j+1}} - \frac{(b\iota\Phi t/\sqrt{5})^j}{\left(1 + \frac{b\iota\Phi t}{\sqrt{5}}\right)^{j+1}} \right],$$

for $t \equiv 2x \Big/ \left(1+\sqrt{1+\frac{4}{5}x^2}\right)$ and where the golden ratio (and its reciprocal) are respectively defined by $\varphi, \Phi := \frac{1}{2}\left(1 \pm \sqrt{5}\right)$.

Inversion Relations and Generating Function Identities

Inversion Relations

An inversion relation is a pair of equations of the form,

$$g_n = \sum_{k=0}^{n} A_{n,k} \cdot f_k \quad \leftrightarrow \quad f_n = \sum_{k=0}^{n} B_{n,k} \cdot g_k,$$

which is equivalent to the orthogonality relation,

$$\sum_{k=j}^{n} A_{n,k} \cdot B_{k,j} = \delta_{n,j}.$$

Given two sequences, $\{f_n\}$ and $\{g_n\}$, related by an inverse relation of the previous form, we sometimes seek to relate the OGFs and EGFs of the pair of sequences by functional equations implied by the inversion relation. This goal in some respects mirrors the more number theoretic (Lambert series) generating function relation guaranteed by the Möbius inversion formula, which provides that whenever,

$$a_n = \sum_{d|n} b_d \quad \leftrightarrow \quad b_n = \sum_{d|n} \mu\left(\frac{n}{d}\right) a_d,$$

the generating functions for the sequences, $\{a_n\}$ and $\{b_n\}$, are related by the Möbius transform given by,

$$\sum_{n\geq 1} a_n z^n = \sum_{n\geq 1} \frac{b_n z^n}{1-z^n}.$$

Similarly, the Euler transform of generating functions for two sequences, $\{a_n\}$ and $\{b_n\}$, satisfying the relation,

$$1 + \sum_{n\geq 1} b_n z^n = \prod_{i\geq 1} \frac{1}{(1-z^i)^{a_i}},$$

is given in the form of,

$$1 + B(z) = \exp\left(\sum_{k\geq 1} \frac{A(z^k)}{k}\right),$$

where the corresponding inversion formulas between the two sequences is given in the reference.

The Binomial Transform

The first inversion relation provided below implicit to the binomial transform is perhaps the simplest of all inversion relations we will consider. For any two sequences, $\{f_n\}$ and $\{g_n\}$, related by the inversion formulas,

$$g_n = \sum_{k=0}^{n} \binom{n}{k} (-1)^k f_k \quad \leftrightarrow \quad f_n = \sum_{k=0}^{n} \binom{n}{k} (-1)^k g_k,$$

we have functional equations between the OGFs and EGFs of these sequences provided by the binomial transform in the forms of,

$$G(z) = \frac{1}{1-z} F\left(\frac{z}{1-z}\right)$$

and

$$\hat{G}(z) = e^z \hat{F}(-z).$$

The Stirling Transform

For any pair of sequences, $\{f_n\}$ and $\{g_n\}$, related by the Stirling number inversion formula,

$$g_n = \sum_{k=1}^{n} \left\{ {n \atop k} \right\} f_k \quad \leftrightarrow \quad f_n = \sum_{k=1}^{n} \left[{n \atop k} \right] (-1)^{n-k} g_k,$$

these inversion relations between the two sequences translate into functional equations between the sequence EGFs given by the Stirling transform as,

$$\hat{G}(z) = \hat{F}\left(e^z - 1\right)$$

and

$$\hat{F}(z) = \hat{G}\left(\log(1+z)\right).$$

ALTERNATING SIGN MATRIX

In mathematics, an alternating sign matrix is a square matrix of 0s, 1s, and −1s such that the sum of each row and column is 1 and the nonzero entries in each row and column alternate in sign. These matrices generalize permutation matrices and arise naturally when using Dodgson condensation to compute a determinant. They are also closely related to the six-vertex model with domain wall boundary conditions from statistical mechanics. They were first defined by William Mills, David Robbins, and Howard Rumsey in the former context.

$$\begin{bmatrix}1&0&0\\0&1&0\\0&0&1\end{bmatrix}\quad\begin{bmatrix}1&0&0\\0&0&1\\0&1&0\end{bmatrix}$$

$$\begin{bmatrix}0&1&0\\1&0&0\\0&0&1\end{bmatrix}\quad\begin{bmatrix}0&1&0\\1&-1&1\\0&1&0\end{bmatrix}\quad\begin{bmatrix}0&1&0\\0&0&1\\1&0&0\end{bmatrix}$$

$$\begin{bmatrix}0&0&1\\1&0&0\\0&1&0\end{bmatrix}\quad\begin{bmatrix}0&0&1\\0&1&0\\1&0&0\end{bmatrix}$$

The seven alternating sign matrices of size 3.

Example:

An example of an alternating sign matrix (that is not also a permutation matrix) is,

$$\begin{bmatrix}0&0&1&0\\1&0&0&0\\0&1&-1&1\\0&0&1&0\end{bmatrix}.$$

Puzzle picture.

Alternating Sign Matrix Conjecture

The alternating sign matrix conjecture states that the number of $n \times n$ alternating sign matrices is

$$\prod_{k=0}^{n-1} \frac{(3k+1)!}{(n+k)!} = \frac{1!4!7!\cdots(3n-2)!}{n!(n+1)!\cdots(2n-1)!}.$$

The first few terms in this sequence for n = 0, 1, 2, 3, ... are,

1, 1, 2, 7, 42, 429, 7436, 218348, ...

This conjecture was first proved by Doron Zeilberger in 1992. In 1995, Greg Kuperberg gave a short proof based on the Yang–Baxter equation for the six-vertex model with domain-wall boundary conditions, that uses a determinant calculation due to Anatoli Izergin.

EULERIAN NUMBER

In combinatorics, the Eulerian number A(n, m) is the number of permutations of the numbers 1 to n in which exactly m elements are greater than the previous element (permutations with m "ascents"). They are the coefficients of the Eulerian polynomials:

$$A_n(t) = \sum_{m=0}^{n} A(n,m)\, t^m.$$

The Eulerian polynomials are defined by the exponential generating function:

$$\sum_{n=0}^{\infty} A_n(t) \cdot \frac{x^n}{n!} = \frac{t-1}{t-e^{(t-1)x}}.$$

The Eulerian polynomials can be computed by the recurrence:

$$A_0(t) = 1,$$

$$A_n(t) = t(1-t) \cdot A'_{n-1}(t) + A_{n-1}(t) \cdot (1+(n-1)t), \quad n \geq 1.$$

An equivalent way to write this definition is to set the Eulerian polynomials inductively by,

$$A_0(t) = 1,$$

$$A_n(t) = \sum_{k=0}^{n-1} \binom{n}{k} A_k(t) \cdot (t-1)^{n-1-k}, \quad n \geq 1.$$

Other notations for A(n, m) are E(n, m) and $\left\langle {n \atop m} \right\rangle$.

Basic Properties

For a given value of n > 0, the index m in A(n, m) can take values from 0 to n – 1. For fixed n there is a single permutation which has 0 ascents: (n, n – 1, n – 2,..., 1). There is also a single permutation which has n – 1 ascents; this is the rising permutation (1, 2, 3,..., n). Therefore A(n, 0) and A(n, n – 1) are 1 for all values of n.

Reversing a permutation with m ascents creates another permutation in which there are n – m – 1 ascents. Therefore A(n, m) = A(n, n – m – 1).

Values of A(n, m) can be calculated "by hand" for small values of n and m. For example:

n	m	Permutations	A(n, m)
1	0	(1)	A(1,0) = 1
2	0	(2, 1)	A(2,0) = 1
	1	(1, 2)	A(2,1) = 1
3	0	(3, 2, 1)	A(3,0) = 1
	1	(1, 3, 2) (2, 1, 3) (2, 3, 1) (3, 1, 2)	A(3,1) = 4
	2	(1, 2, 3)	A(3,2) = 1

For larger values of n, A(n, m) can be calculated using the recursive formula:

$$A(n,m) = (n-m)A(n-1,m-1) + (m+1)A(n-1,m).$$

For example:

$$A(4,1) = (4-1)A(3,0) + (1+1)A(3,1) = 3\times1 + 2\times4 = 11.$$

Values of A(n, m) for $0 \le n \le 9$ are:

m/n	0	1	2	3	4	5	6	7	8
1	1								
2	1	1							
3	1	4	1						
4	1	11	11	1					
5	1	26	66	26	1				
6	1	57	302	302	57	1			
7	1	120	1191	2416	1191	120	1		
8	1	247	4293	15619	15619	4293	247	1	
9	1	502	14608	88234	156190	88234	14608	502	1

The above triangular array is called the Euler triangle or Euler's triangle, and it shares some common characteristics with Pascal's triangle. The sum of row n is the factorial n!.

Explicit Formula

An explicit formula for A(n, m) is,

$$A(n,m)=\sum_{k=0}^{m}(-1)^{k}\binom{n+1}{k}(m+1-k)^{n}.$$

One can see from this formula, as well as from the combinatorial interpretation, that $A(n,n)=0$ for n 0,, so that $A_n(t)$ is a polynomial of degree $n-1$ for $n>0$.

Summation Properties

It is clear from the combinatorial definition that the sum of the Eulerian numbers for a fixed value of n is the total number of permutations of the numbers 1 to n, so,

$$\sum_{m=0}^{n-1}A(n,m)=n! \text{ for } n\geq 1.$$

The alternating sum of the Eulerian numbers for a fixed value of n is related to the Bernoulli number B_{n+1},

$$\sum_{m=0}^{n-1}(-1)^{m}A(n,m)=\frac{2^{n+1}(2^{n+1}-1)B_{n+1}}{n+1} \text{ for } n\geq 1.$$

Other summation properties of the Eulerian numbers are:

$$\sum_{m=0}^{n-1}(-1)^{m}\frac{A(n,m)}{\binom{n-1}{m}}=0 \text{ for } n\geq 2,$$

$$\sum_{m=0}^{n-1}(-1)^{m}\frac{A(n,m)}{\binom{n}{m}}=(n+1)B_{n} \text{ for } n\geq 2,$$

where B_n is the nth Bernoulli number.

Identities

The Eulerian numbers are involved in the generating function for the sequence of nth powers:

$$\sum_{k=0}^{\infty}k^{n}x^{k}=\frac{\sum_{m=0}^{n-1}A(n,m)x^{m+1}}{(1-x)^{n+1}}=\frac{x\cdot A_{n}(x)}{(1-x)^{n+1}}$$

for $n \geq 0$. This assumes that $0^0 = 0$ and $A(0,0) = 1$ (since there is one permutation of no elements, and it has no ascents).

Worpitzky's identity expresses x^n as the linear combination of Eulerian numbers with binomial coefficients:

$$x^n = \sum_{m=0}^{n-1} A(n,m)\binom{x+m}{n}.$$

It follows from Worpitzky's identity that,

$$\sum_{k=1}^{N} k^n = \sum_{m=0}^{n-1} A(n,m)\binom{N+1+m}{n+1}.$$

Another interesting identity is,

$$\frac{e}{1-ex} = \sum_{n=0}^{\infty} \frac{A_n(x)}{n!(1-x)^{n+1}}.$$

The numerator on the right-hand side is the Eulerian polynomial.

For a fixed function $f:\mathbb{R} \to \mathbb{C}$ which is integrable on $(0,n)$ we have the integral formula:

$$\int_0^1 \cdots \int_0^1 f(\lfloor x_1 + \cdots + x_n \rfloor) dx_1 \cdots dx_n = \sum_k A(n,k)\frac{f(k)}{n!}.$$

Eulerian Numbers of the Second Kind

The permutations of the multiset {1, 1, 2, 2, ⋯, n, n} which have the property that for each k, all the numbers appearing between the two occurrences of k in the permutation are greater than k are counted by the double factorial number (2n–1)!!. The Eulerian number of the second kind, denoted $\left\langle\!\left\langle {n \atop m} \right\rangle\!\right\rangle$, counts the number of all such permutations that have exactly m ascents. For instance, for n = 3 there are 15 such permutations, 1 with no ascents, 8 with a single ascent, and 6 with two ascents:

332211,

221133, 221331, 223311, 233211, 113322, 133221, 331122,

331221,

112233, 122133, 112332, 123321, 133122, 122331.

The Eulerian numbers of the second kind satisfy the recurrence relation, that follows directly from the above definition:

$$\left\langle\!\left\langle {n \atop m} \right\rangle\!\right\rangle = (2n-m-1)\left\langle\!\left\langle {n-1 \atop m-1} \right\rangle\!\right\rangle + (m+1)\left\langle\!\left\langle {n-1 \atop m} \right\rangle\!\right\rangle,$$

with initial condition for n = 0, expressed in Iverson bracket notation:

$$\left\langle\!\left\langle {0 \atop m} \right\rangle\!\right\rangle = [m=0].$$

Correspondingly, the Eulerian polynomial of second kind, here denoted P_n (no standard notation exists for them) are:

$$P_n(x) := \sum_{m=0}^{n} \left\langle\!\left\langle {n \atop m} \right\rangle\!\right\rangle x^m$$

and the above recurrence relations are translated into a recurrence relation for the sequence $P_n(x)$:

$$P_{n+1}(x) = (2nx+1)P_n(x) - x(x-1)P_n'(x)$$

with initial condition,

$$P_0(x) = 1.$$

The latter recurrence may be written in a somehow more compact form by means of an integrating factor:

$$(x-1)^{-2n-2}P_{n+1}(x) = \left(x(1-x)^{-2n-1}P_n(x)\right)'$$

so that the rational function,

$$u_n(x) := (x-1)^{-2n}P_n(x)$$

satisfies a simple autonomous recurrence:

$$u_{n+1} = \left(\frac{x}{1-x}u_n\right)', \quad u_0 = 1,$$

whence one obtains the Eulerian polynomials as $P_n(x) = (1-x)^{2n} u_n(x)$, and the Eulerian numbers of the second kind as their coefficients.

Here are some values of the second order Eulerian numbers:

m/n	0	1	2	3	4	5	6	7	8
1	1								
2	1	2							
3	1	8	6						
4	1	22	58	24					
5	1	52	328	444	120				
6	1	114	1452	4400	3708	720			
7	1	240	5610	32120	58140	33984	5040		
8	1	494	19950	195800	644020	785304	341136	40320	
9	1	1004	67260	1062500	5765500	12440064	11026296	3733920	362880

The sum of the n-th row, which is also the value $P_n(1)$, is $(2n-1)!!$.

EXPONENTIAL FORMULA

In combinatorial mathematics, the exponential formula (called the polymer expansion in physics) states that the exponential generating function for structures on finite sets is the exponential of the exponential generating function for connected structures. The exponential formula is a power-series version of a special case of Faà di Bruno's formula.

For any formal power series of the form:

$$f(x) = a_1 x + \frac{a_2}{2} x^2 + \frac{a_3}{6} x^3 + \cdots + \frac{a_n}{n!} x^n + \cdots$$

we have,

$$\exp f(x) = e^{f(x)} = \sum_{n=0}^{\infty} \frac{b_n}{n!} x^n,$$

where,

$$b_n = \sum_{\pi=\{S_1,\ldots,S_k\}} a_{|S_1|} \cdots a_{|S_k|},$$

and the index π runs through the list of all partitions $\{ S_1,\ldots, S_k \}$ of the set $\{ 1,\ldots, n \}$. (When $k = 0$, the product is empty and by definition equals 1.)

One can write the formula in the following form:

$$b_n = B_n(a_1, a_2, \ldots, a_n),$$

and thus,

$$\exp\left(\sum_{n=1}^{\infty} \frac{a_n}{n!} x^n\right) = \sum_{n=0}^{\infty} \frac{B_n(a_1, \ldots, a_n)}{n!} x^n,$$

where $B_n(a_1, \ldots, a_n)$ is the nth complete Bell polynomial.

Examples:

- $b_3 = B_3(a_1, a_2, a_3) = a_3 + 3a_2a_1 + a_1^3$, because there is one partition of the set { 1, 2, 3 } that has a single block of size 3, there are three partitions of { 1, 2, 3 } that split it into a block of size 2 and a block of size 1, and there is one partition of { 1, 2, 3 } that splits it into three blocks of size 1.
- If $b_n = 2^{n(n-1)/2}$ is the number of graphs whose vertices are a given n-point set, then a_n is the number of connected graphs whose vertices are a given n-point set.
- There are numerous variations of the previous example where the graph has certain properties: for example, if b_n counts graphs without cycles, then a_n counts trees (connected graphs without cycles).
- If b_n counts directed graphs whose edges (rather than vertices) are a given n point set, then a_n counts connected directed graphs with this edge s

Applications

In applications, the numbers a_n often count the number of some sort of "connected" structure on an n-point set, and the numbers b_n count the number of (possibly disconnected) structures. The numbers $b_n/n!$ count the number of isomorphism classes of structures on n points, with each structure being weighted by the reciprocal of its automorphism group, and the numbers $a_n/n!$ count isomorphism classes of connected structures in the same way.

In quantum field theory and statistical mechanics, the partition functions Z, or more generally correlation functions, are given by a formal sum over Feynman diagrams. The exponential formula shows that log(Z) can be written as a sum over connected Feynman diagrams, in terms of connected correlation functions.

GRAPH ENUMERATION

In combinatorics, an area of mathematics, graph enumeration describes a class of combinatorial enumeration problems in which one must count undirected or directed graphs of certain types, typically as a function of the number of vertices of the graph. These problems may be solved either exactly (as an algebraic enumeration problem) or asymptotically. The pioneers in this area of mathematics were George Pólya, Arthur Cayley and John Howard Redfield.

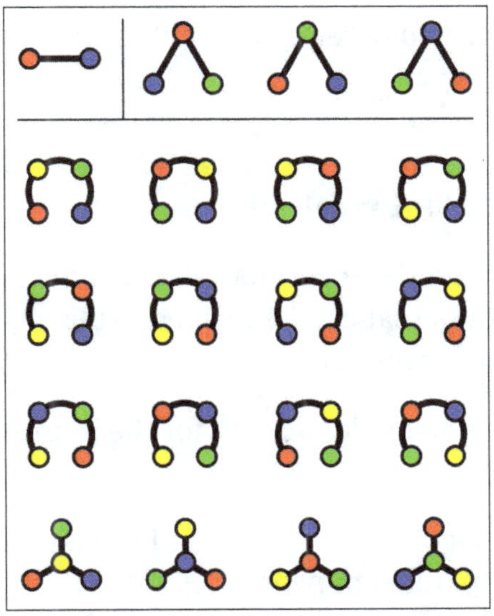

The complete list of all free trees on 2,3,4 labeled vertices: $2^{2-2}=1$ tree with 2 vertices, $3^{3-2}=3$ trees with 3 vertices and $4^{4-2}=16$ trees with 4 vertices.

Labeled vs. Unlabeled Problems

In some graphical enumeration problems, the vertices of the graph are considered to be labeled in such a way as to be distinguishable from each other, while in other problems any permutation of the vertices is considered to form the same graph, so the vertices are considered identical or unlabeled. In general, labeled problems tend to be easier. As with combinatorial enumeration more generally, the Pólya enumeration theorem is an important tool for reducing unlabeled problems to labeled ones: each unlabeled class is considered as a symmetry class of labeled objects.

Exact Enumeration Formulas

Some important results in this area include the following.

- The number of labeled n-vertex simple undirected graphs is $2^{n(n-1)/2}$.
- The number of labeled n-vertex simple directed graphs is $2^{n(n-1)}$.
- The number C_n of connected labeled n-vertex undirected graphs satisfies the recurrence relation.

$$C_n = 2^{\binom{n}{2}} - \frac{1}{n}\sum_{k=1}^{n-1} k\binom{n}{k} 2^{\binom{n-k}{2}} C_k.$$

from which one may easily calculate, for n = 1, 2, 3,..., that the values for C_n are,

1, 1, 4, 38, 728, 26704, 1866256.

- The number of labeled n-vertex free trees is n^{n-2} (Cayley's formula).
- The number of unlabeled n-vertex caterpillars is,

$$2^{n-4} + 2^{\lfloor (n-4)/2 \rfloor}.$$

Series-parallel Networks Problem

In combinatorial mathematics, the series-parallel networks problem asks for the number of series-parallel networks that can be formed using a given number of edges. The edges can be distinguishable or indistinguishable.

When the edges are indistinguishable, we look for the number of topologically different networks on n edges.

Solution: The idea is to break-down the problem by classifying the networks as essentially series and essentially parallel networks.

- An essentially series network is a network which can be broken down into two or more subnetworks in series.
- An essentially parallel network is a network which can be broken down into two or more subnetworks in parallel.

By the duality of networks, it can be proved that the number of essentially series networks is equal to the number of essentially parallel networks. Thus for all n > 1, the number of networks in n edges is twice the number of essentially series networks. For n = 1, the number of networks is 1.

Define:

- a_n as the number of series-parallel networks on n indistinguishable edges.
- b_n as the number of essentially series networks.

Then,

$$a_n = \begin{cases} 1, & \text{if n is 1} \\ 2b_n, & \text{otherwise} \end{cases}$$

b_n can be found out by enumerating the partitions of n.

Consider a partition, $\{p_i\}$ of n:

$$\sum_i ip_i = n.$$

Consider the essentially series networks whose components correspond to the partition above. That is the number of components with i edges is p_i. The number of such networks can be computed as,

$$\prod_i \binom{b_i + p_i - 1}{p_i}.$$

Hence,

$$b_n = \sum_{p_i} \prod_i \binom{b_i + p_i - 1}{p_i}$$

where the summation is over all partitions, p_i of n except for the trivial partition consisting of only n.

This gives a recurrence for computing b_n. Now a_n can be computed as above.

Wedderburn–Etherington Number

The Wedderburn–Etherington numbers are an integer sequence named for Ivor Malcolm Haddon Etherington and Joseph Wedderburn that can be used to count certain kinds of binary trees. The first few numbers in the sequence are:

0, 1, 1, 1, 2, 3, 6, 11, 23, 46, 98, 207, 451, 983, 2179, 4850, 10905, 24631, 56011,...

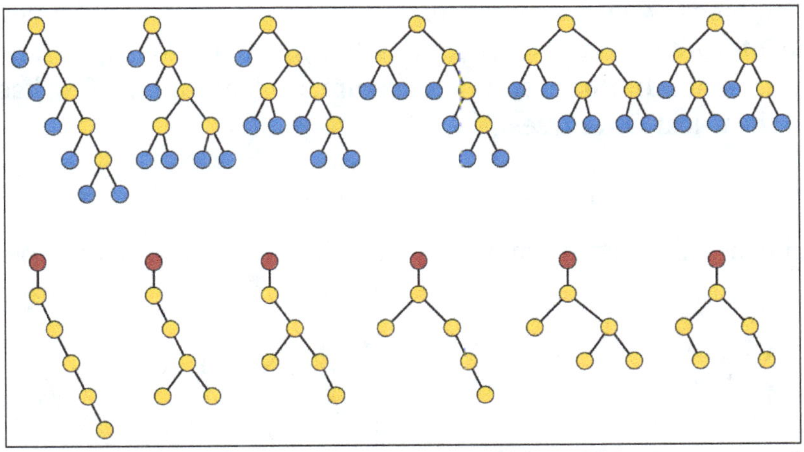

Otter trees and weakly binary trees, two types of rooted binary tree counted by the Wedderburn–Etherington numbers.

These numbers can be used to solve several problems in combinatorial enumeration. The nth number in the sequence (starting with the number 0 for n = 0) counts:

- The number of unordered rooted trees with n leaves in which all nodes including the root have either zero or exactly two children. These trees have been called Otter trees, after the work of Richard Otter on their combinatorial enumeration.

They can also be interpreted as unlabeled and unranked dendrograms with the given number of leaves.

- The number of unordered rooted trees with n nodes in which the root has degree zero or one and all other nodes have at most two children. Trees in which the root has at most one child are called planted trees, and the additional condition that the other nodes have at most two children defines the weakly binary trees. In chemical graph theory, these trees can be interpreted as isomers of polyenes with a designated leaf atom chosen as the root.
- The number of different ways of organizing a single-elimination tournament for n players (with the player names left blank, prior to seeding players into the tournament). The pairings of such a tournament may be described by an Otter tree.
- The number of different results that could be generated by different ways of grouping the expression x^n for a binary multiplication operation that is assumed to be commutative but neither associative nor idempotent. For instance x^5 can be grouped into binary multiplications in three ways, as $x(x(x(xx)))$, $x((xx)(xx))$, or $(xx)(x(xx))$. This was the interpretation originally considered by both Etherington and Wedderburn. An Otter tree can be interpreted as a grouped expression in which each leaf node corresponds to one of the copies of x and each non-leaf node corresponds to a multiplication operation. In the other direction, the set of all Otter trees, with a binary multiplication operation that combines two trees by making them the two subtrees of a new root node, can be interpreted as the free commutative magma on one generator x (the tree with one node). In this algebraic structure, each grouping of x^n has as its value one of the n-leaf Otter trees.

Formula:

The Wedderburn–Etherington numbers may be calculated using the recurrence relation,

$$a_{2n-1} = \sum_{i=1}^{n-1} a_i a_{2n-i-1}$$

beginning with the base case $a_1 = 1$.

In terms of the interpretation of these numbers as counting rooted binary trees with n leaves, the summation in the recurrence counts the different ways of partitioning these leaves into two subsets, and of forming a subtree having each subset as its leaves. The formula for even values of n is slightly more complicated than the formula for odd values in order to avoid double counting trees with the same number of leaves in both subtrees.

Growth Rate

The Wedderburn–Etherington numbers grow asymptotically as,

$$a_n \approx \sqrt{\frac{\rho+\rho^2 B'(\rho^2)}{2\pi}} \frac{\rho^{-n}}{n^{3/2}},$$

where B is the generating function of the numbers and ρ is its radius of convergence, approximately 0.4027, and where the constant given by the part of the expression in the square root is approximately 0.3188.

Applications

Young & Yung use the Wedderburn–Etherington numbers as part of a design for an encryption system containing a hidden backdoor. When an input to be encrypted by their system can be sufficiently compressed by Huffman coding, it is replaced by the compressed form together with additional information that leaks key data to the attacker. In this system, the shape of the Huffman coding tree is described as an Otter tree and encoded as a binary number in the interval from 0 to the Wedderburn–Etherington number for the number of symbols in the code. In this way, the encoding uses a very small number of bits, the base-2 logarithm of the Wedderburn–Etherington number.

Farzan & Munro describe a similar encoding technique for rooted unordered binary trees, based on partitioning the trees into small subtrees and encoding each subtree as a number bounded by the Wedderburn–Etherington number for its size. Their scheme allows these trees to be encoded in a number of bits that is close to the information-theoretic lower bound (the base-2 logarithm of the Wedderburn–Etherington number) while still allowing constant-time navigation operations within the tree. Iserles & Nørsett use unordered binary trees, and the fact that the Wedderburn–Etherington numbers are significantly smaller than the numbers that count ordered binary trees, to significantly reduce the number of terms in a series representation of the solution to certain differential equations.

LATTICE PATH

A lattice path (path for short) is a path (walk) in a lattice in some d-dimensional Euclidean space.

A lattice path (path for short) is what the name says: a path (walk) in a lattice in some d-dimensional Euclidean space. Formally, a lattice path P is a sequence $P = (P_0, P_1, \ldots, P_l)$ of points Pi in $\mathbb{Z}^d$. Figure shows the lattice path ((0,0),(1,1), (2,1),(3,1),(3,2),(4,3)). The point Po is called the starting point and Pl is called the end point of P. The vectors $\overrightarrow{P_0P_1}, \overrightarrow{P_1P_2}, \ldots, \overrightarrow{P_{l-1}P_l}$ are called the steps of P.

Lattice paths have been studied for a very long time, explicitly at least since the second half of the 19th century. At the beginning stand the investigations concerning the two-candidate ballot problem and the gambler's ruin problem. Since then, lattice paths have penetrated many fields of mathematics, computer science, and physics. The reason for their ubiquity is, on the one hand, that they are well-suited to encode various (combinatorial) objects and their properties, and, thus, problems in various fields can be solved by solving lattice path problems. On the other hand, since lattice paths are — at the outset — reasonably simple combinatorial objects, the study of physical, probabilistic, or statistical models is attractive in its own right.

References

- Goulden, Ian P.; Jackson, David M. (2004). Combinatorial Enumeration. Dover Publications. ISBN 978-0486435978
- Bóna, Miklós; Flajolet, Philippe (2009), "Isomorphism and symmetries in random phylogenetic trees", Journal of Applied Probability, 46 (4): 1005–1019, arXiv:0901.0696, Bibcode:2009arXiv0901.0696B, doi:10.1239/jap/1261670685, MR 2582703
- Flajolet, Philippe; Sedgewick, Robert (2009). Analytic Combinatorics. Cambridge University Press. ISBN 978-0-521-89806-5. Zbl 1165.05001
- Schmidt, M. D. (2017). "Jacobi-Type Continued Fractions for the Ordinary Generating Functions of Generalized Factorial Functions". Journal of Integer Sequences
- Petersen, T. Kyle (2015). Eulerian Numbers. Birkhäuser. pp. 3–18. doi:10.1007/978-1-4939-3091-3_1

Additive Combinatorics

CHAPTER 5

Additive combinatorics is a special case of combinatorics that only uses the operations of addition and subtraction. It further includes Ruzsa triangle inequality, Gowers norm, sum-free sequence, restricted sumset, etc. This chapter delves into the subject of additive combinatorics for a thorough understanding of it.

Additive combinatorics is an active branch of mathematics that interfaces with combinatorics, number theory, ergodic theory, harmonic analysis and geometry over finite fields.

Let A and B be subsets of G, an additive group. Typically we work with the integers Z, or the integers mod N, that is ($\mathbb{Z}/N\mathbb{Z}$), though sometimes with other groups like $\mathbb{R}$ or $\mathbb{Z}^k$. The sumset of A and B is defined by,

$$A+B := \{g \in G : \text{There exist } a \in A,\ b \in B \text{ such that } g = a+b\}.$$

Typically we write $A+B=\{a+b:\ a\in A, b\in B\}$ with the understanding that elements are not repeated in A + B. For example, {1, 2, 3} + {1, 3} = {2, 3, 4, 5, 6}. The addition of sets, "+", is commutative if (G, +) is commutative. It is also associative, and it is distributive over unions, that is, A + (B ∪ C) = (A + B) ∪ (A + C).

Other important definitions include:

$$\begin{aligned} kA &:= A + A + \cdots + A; \\ b + A &= \{b\} + A, \text{ a translate of } A; \\ A - B &= \{a - b : a \in A, b \in B\}; \\ k \Diamond A &= \{ka : a \in A\}, \text{ a dilate of } A; \\ \text{and } A \Diamond B &= \{ab : a \in A, b \in B\}. \end{aligned}$$

Having given all this notation we note that we will abuse it by writing $N\mathbb{Z}$ instead of $N \Diamond \mathbb{Z}$, for the integers divisible by N.

If A + B is Small then A and B Are...?

Suppose that A and B are finite sets of integers, say A is $a_1 < a_2 < \cdots < a_r$, and B is $b_1 < b_2 < \cdots < b_s$. Then A + B contains the r + s − 1 distinct elements:

$$a_1+b_1 < a_1+b_2 < a_1+b_3 < \cdots < a_1+b_s < a_2+b_s < \cdots < a_r+b_s,$$

so that,

$$|A+B| \geq |A|+|B|-1.$$

Can we have equality in the above equation? That is, what if $|A+B|=|A|+|B|-1$? We will write down another list r + s – 1 distinct elements of A + B, namely,

$$a_1+b_1 < a_2+b_1 < a_2+b_2 < \cdots < a_2+b_{s-1} < a_2+b_s < \cdots < a_r+b_s.$$

If $|A+B|=r+s-1$, then the terms in each list must be the same and so we have $a_1+b_2=a_2+b_1$, and $a_1+b_3=a_2+b_2$, etc., implying that $a_2-a_1=b_2-b_1=b_3-b_2=\dots$. In fact we can deduce that A and B are both arithmetic progressions with the same common difference; that is there exists a non-zero integer d such that,

$$A=\{a+id : 0 \leq i \leq I-1\} \text{ and } B=\{b+jd : 0 \leq j \leq J-1\}.$$

Thus A and B are highly structured. However if A is a large subset of $\{a+id : 0 \leq i \leq I-1\}$ and B is a large subset of $\{b+jd : 0 \leq j \leq J-1\}$, then we expect that $|A+B|=|A|+|B|+\Delta$ for some small Δ, yet A and B may not have much internal structure. The key thing is that they are both large subsets of arithmetic progressions with the same common difference. Another interesting case is given by,

$$\begin{aligned} A &= \{1, 2, \dots, 10, 101, 102, \dots, 110, 201, 202, \dots, 210\} \\ &= 1+\{0, 1, \dots, 9\} + 100 \lozenge \{0, 1, 2\}, \\ B &= 3+\{0, 1, \dots, 7\}+100 \lozenge \{0, 1, \dots, 4\}, \\ \text{and } A+B &= 4+\{0, 1, 2, \dots, 16\} + 100 \lozenge \{0, 1, \dots, 6\}, \end{aligned}$$

so that $|A|=30, |B|=40$ and $|A+B|=119$. These are examples of a generalized arithmetic progression (GAP):

$$C := \{a_0+a_1n_1+a_2n_2+\cdots+a_kn_k : 0 \leq n_j \leq N_j-1 \text{ for } 1 \leq j \leq k\},$$

where $N_1, N_2, \dots, N_k$ are integers ≥ 2. This GAP is said to have dimension k and volume $N_1N_2\dots Nk$; and is proper if its elements are distinct.

Most questions about the structure of A and B, when $A+B$ is small, are open. We study the structure of A when $A+A=2A$ is small (i.e., the case B = A). For a GAP C we have $|2C|<2^k|C|$; and, indeed, if $A \subset C$ with $|A| \geq \delta|C|$ then,

$$|2A| \leq |2C| < 2^k|C| \leq (2^k/\delta)|A|.$$

If $|2A|$ is a small multiple of $|A|$ then what possible A are there? A rather daring guess is that the only possible such A are large subsets of GAPs.

The Freiman-Ruzsa Theorem

If $|2A|$ is "small" then A is a "large" subset of a GAP.

Densities

The Schnirelmann density of a set A of integers is given by,

$$\sigma(A) := \inf_{n\geq 1} \frac{\#\{a \in A : 1 \leq a \leq n\}}{n},$$

so that $A(n) \geq n\sigma(A)$ for all $n \geq 1$. It is easy to see, by the pigeonhole principle that if $0 \in A \cap B$ and $\sigma(A)+\sigma(B) \geq 1$ then $A + B \supseteq \mathbb{Z}_{\geq 0}$. By counting the elements in $A+B$ of the form $a_i + b_j$ with $ai \leq a_i + b_j < a_{i+1}$, Schnirelmann proved that if $1 \in A$ and $0 \in B$ then,

$$\sigma(A + B) \geq \sigma(A) + \sigma(B) - \sigma(A)\sigma(B).$$

This is more usefully rewritten as $(1-\sigma(A+B)) \leq (1-\sigma(A))(1-\sigma(B))$ since then we see that $(1-\sigma(hA)) \leq (1-\sigma(A))^h$. The last two results thus imply that if $1 \in A$ and $\sigma(A) > 0$ then A is a basis of order 2h where the integer h is chosen so that $(1-\sigma(A))^h \leq 1/2$. (Note that $\sigma(A) > 0$ implies that $1 \in A$; and that some condition like $1 \in A$ is necessary to avoid A, and hence hA, being a subset of the even integers.)

The lower density $\underline{d}(A)$ is defined by:

$$\underline{d}(A) := \liminf_{n\to\infty} \frac{\#\{a \in A : 1 \leq a \leq n\}}{n}$$

We will prove that if $\underline{d}(A) > 0$ then for all $\varepsilon \in (0, \underline{d}(A))$ there exists $r = r_a$ such that $\sigma\left(A_{(r)}\right) \geq \underline{d}(A) - \epsilon$, where $A_{(r)} = \{a - r : a \in A,\ a > r\}$. There exists an integer n_ε such that if $n \geq n_a$ then $\#\{a \in A : 1 \leq a \leq n\} \geq (\underline{d}(A) - ^a)n$.

If there exists any $n \geq n^2$ with $\#\{a \in A : 1 \leq a \leq n\} < \underline{d}(A)n$ then there must be an $n \geq n^a$, say $n = m^a$, with $\rho^a := \#\{a \in A : 1 \leq a \leq n\}/n$ minimal. Hence if $n > m^a$ then,

$$\#\{a \in A : m^a < a \leq n\} \geq \rho^a (n - m^a) \geq (\underline{d}(A) - ^a)(n - m^a).$$

On the other hand if $\#\{a \in A : 1 \leq a \leq n\} \geq d(A)n$ for all $n \geq n^a$ then either $\sigma(A) \geq d(A)$, or there exists a maximal r^a (which is necessarily $< n^a$) with $\#\{a \in A : 1 \leq a \leq r^a\} < \underline{d}(A)r^a$, and the result follows:

A straightforward sieve argument implies that at least 1/4 of the even integers can be written as the sum of two primes; that is $\underline{d}(2\mathbb{P} \geq 3) \geq 1/8$. Using the argument of the previous paragraph, and Schnirelmann's theorem, one can prove that the primes are a basis of order 11 (or less). It can also be shown that the k-th powers of integers form an additive basis.

For a finite set of integers S define the cube $\overline{S}$ by,

$$\overline{S} := \left\{ \sum_{s \in S} \varepsilon_s s : \varepsilon_s \in \{-1, 0, 1\} \text{ for all } s \in S, \right\}$$

which is a GAP of dimension |S| and volume 3|S|.

Theorem: If A is a set of integers with $\underline{d}(A) > 0$, then there exists a finite set of integers S such that A – A + $\overline{S}$ = $\mathbb{Z}$.

Proof: If A – A 6= Z then there exists $m \notin A - A$, and so A and m + A are disjoint. Let $A_1 = A \cup (m + A)$, so that $d(A_1) = 2\underline{d}(A)$ and $A_1 - A_1 = A - A + \{m\}$. If this is not $\mathbb{Z}$, define $A_2, A_3, \ldots$ Therefore $|S| \leq k$ where k is the largest integer for which $2^k \underline{d}(A) \leq 1$.

Since $\underline{d}(2\mathbb{P}_{\geq 3}) \geq 1/8$, we can deduce that there exists a set S1 of no more than three integers for which,

$$\mathbb{Z} = 2\mathbb{P}_{\geq 3} - 2\mathbb{P}_{\geq 3} + \overline{S_1}$$

It is interesting to determine how small a set one needs to "complete" a given set in this manner. Thus above we added $\overline{S_1}$ to $2\mathbb{P}_{\geq 3} - 2\mathbb{P}_{\geq 3}$ to obtain Z, though we believe that $\mathbb{P}_{\geq 3}\mathbb{P}_{\geq 3} + \{0, 1\} = \mathbb{Z}$. For sums of squares we have $4\{n^2 : n \in \mathbb{Z}\} =_{\geq 0}$; and one can show that $3\{n^2 : n \in Z\} + \{0, 2\} = \mathbb{Z}_{\geq 0}$. A challenge is to find "thin" sets B and C for which $2\{n^2 : n \in \mathbb{Z}\} + B = \mathbb{Z}_{\geq 0}$, and for which $\mathbb{P} + C = \mathbb{Z} \geq 0$.

The Dyson Transformation

However, once Freeman Dyson introduced a simple map between pairs of sets, found new, cleaner arguments in many of the essential questions: For,

$e \in A$ let Be := $\{b \in B : b + e \notin A\}$, and define the Dyson transformation of A, B with respect to e to be,

$$\delta_e(A) := A \cup (e + B) = A \cup (e + B_e), \text{ and } \delta_e(B) := B \setminus B_e.$$

Notice that Be $\subseteq B$ and $(e + B_e) \cap A = \varnothing$. There are several other observations to be made besides:

$$e + \delta_e(B) \subseteq A \subseteq \delta_e(A), \text{ and } |\delta_e(A)| + |\delta_e(B)| = |A| + |B|;$$

$$A \cap (e + B) = e + \delta_e(B) = \delta_e(A) \cap (e + \delta_e(B)),$$

$$\text{and } A \cup (e + B) = \delta_e(A) = \delta_e(A) \cup (e + \delta_e(B)),$$

$$\text{as well as the non-trivial } \delta_e(A) + \delta_e(B) \subseteq A + B.$$

Using a sequence of Dyson transformations one can easily prove Mann's "best possible" improvement of Schnirelmann's theorem.

Mann's Theorem

If $0 \in A \cap B$ then,

$$\sigma(A + B) \geq \min\{1, \sigma(A) + \sigma(B)\}.$$

Note that this result does not extend directly to questions about lower density; that is, $\underline{d}(A+B) \geq \min\{1, \underline{d}(A)+\underline{d}(B)\}$ is not true in general: For example, if,

$$A = B = \{n \equiv 0 \text{ or } 1 \pmod{m}\} \text{ then } A + B = \{n \equiv 0, 1 \text{ or } 2 \pmod{m}\}.$$

So, to understand set addition with respect to lower density, we certainly need to understand set addition mod N.

The Cauchy-Davenport Theorem

If A and B are non-empty subsets of $\mathbb{Z}/N\mathbb{Z}$ where $0 \in B$, and $(b, N) = 1$ for all $b \in B \setminus \{0\}$ then,

$$|A + B| \geq \min\{N, |A| + |B| - 1\}.$$

Proof: By induction on $|B|$: If $|B| = 1$ then $B = \{0\}$ so $A + B = A$ which is okay. We may assume that $1 \leq |A| \leq N - 1$. Now $A + B \neq A$ else for each $b \in B$, for all $a \in A$ there exists a $0 \in A$ such that $a + b \equiv a'$ (mod N). Running through all $a \in A$ we obtain all $a\ 0 \in A$, and so taking the sum over all $a \in A$ we get $|A|b \equiv 0 \pmod{N}$. By selecting non-zero $b \in B$ we have $(b, N) = 1$, and so N divides |A|, which is impossible.

So take $e \in A$ for which $e + b \ / \in A$. By the induction hypothesis the result holds for the pair $\delta_e(A)$, $\delta_e(B)$ (which are non-empty since $A \subseteq \delta_e(A)$ and $0 \in \delta_e(B)$), so that,

$$|A+B| \geq |\delta_e(A)+\delta_e(B)| \geq \min\{N, |\delta_e(A)| + |\delta_e(B)| - 1\} = \min\{N, |A| + |B| - 1\}.$$

Corollary: If $A, B \subseteq \mathbb{Z}/p\mathbb{Z}$ with p prime then $|A + B| \geq \min\{p, |A| + |B| - 1\}$.

There are just three cases in which we get equality (that is, $|A+B| = |A| + |B| - 1$) when $A + B$ is a proper subset of $\mathbb{Z}/p\mathbb{Z}$:

- Either A or B has just one element (that is, $|A| = 1$ or $|B| = 1$).
- A and B are segments of arithmetic progressions with the same common difference (that is, $A = a + d\Diamond\{0, 1, \ldots, r - 1\}$, and $B = b + d\Diamond\{0, 1, \ldots, s - 1\}$ for some $r + s \leq p$).
- A and B are selected maximally so that $d \notin A + B$ (that is, $A \cup (d - B)$ is a partition of $\mathbb{Z}/p\mathbb{Z}$ for some integer d).

Simple Inequalities for Sizes of Sumsets

The Freiman-Ruzsa theorem tells us that if $|A+A|<C|A|$ then A is a large subset of a d-dimensional GAP, G, for some d that can be bounded as a function of C. This implies that A – A is a large subset of G – G, a GAP that is at most twice as large (in each direction) as G, and so $|A - A| \le 2^d |G| \le 2^d C' |A|$ for some constant C′ which depends only on C. Similarly kA – ℓ A is a large subset of kG – ℓ G, also a d-dimensional GAP, and so $|kA - \ell A| \le (k+\ell)^d C'|A|$.

Here, we derive consequences of this type direct without using the relatively deep Freiman-Ruzsa theorem; that is, our objective is to prove that if $|A+A|<C|A|$ then $|kA - \ell A| \le Ck,\ell|A|$ for some constant $C_{k,\ell}$ which depends only on C, k, ℓ. We will see that there are several easy approaches to this problem. When we prove the Freiman-Ruzsa theorem, we will use such inequalities in our proof. We start with the most basic question of this type:

The relationship between $A+A$ and $A-A$. We prove that,

$$\frac{1}{2} \le \log\left(\frac{|A+A|}{|A|}\right) / \log\left(\frac{|A-A|}{|A|}\right) \le 3;$$

we are interested in determining the strongest possible form of each of these inequalities. We give two examples:

For $A = \{0, 1, 3\}$, we have $A+A=\{0, 1, 2, 3, 4, 6\}$ and $A-A=\{-3,-2,-1,0,1,2,3\}$, so that $|A+A|=6<|A-A|=7$.

- For $A=\{0,2,3,4,7,11,12,14\}$ we have $A+A=[0, 28]\cap\mathbb{Z}\setminus\{1,20,27\}$ and,

  $A-A=([-14, 14]\cap\mathbb{Z})\setminus\{-13,-6,6,13\}$, so that $|A-A|=25<|A+A|=26$.

These isolated examples can be made into arbitrarily large examples by using the Cartesian product: The idea simply is to take $B=A^{(k)}=A\times\cdot\cdot\times A$, so in the first case $|B+B|=6^k<|B-B|=7^k$. One might object that B is not a subset of the integers but in fact the bijection $B \leftrightarrow C$ defined by (ao,..., ak−1) $(a_0,\ldots,a_{k-1})a_0+a_1 7+\cdots+a_{k-1}7^{k-1}$ is also a bijection, when correctly interpreted, between the sets $B+B$ and $C+C$, and between B – B and C – C. This map is called a Freiman 2-isomorphism (the "2" since it remains a bijection when we add two elements of our set).

we will discuss this in detail in our next lecture. We thus conclude from our examples that the constant "$\frac{1}{2}$" in $\frac{1}{2} \le \log\left(\frac{|A+A|}{|A|}\right) / \log\left(\frac{|A-A|}{|A|}\right) \le 3$; may not be increased beyond $\frac{\log(6/3)}{\log(7/3)} = .81806\ldots$; and the constant "3" in $\frac{1}{2} \le \log\left(\frac{|A+A|}{|A|}\right) / \log\left(\frac{|A-A|}{|A|}\right) \le 3$; may not be decreased beyond $\frac{\log(26/8)}{\log(25/8)} = 1.03442\ldots$ A better example for the lower bound comes from taking $A=\{1, 2, 2^2,...,2\ 5\}$ so that $|A|=6, |A+A|=21<|A-A|=31$ as $\frac{\log(21/6)}{\log(31/6)} = .762843\ldots$

Some First Bounds

We begin by establishing that for any finite sets A, B, C inside an additive group G (whether commutative or not) we have,

$$|A - C|\ |B| \le |A - B|\ |B - C|,$$

by showing that there is an injection $\phi:(A-C)\times B \to (A-B)\times(B-C)$: For each $\lambda \in A-C$ fix $a_\lambda \in A$, $c_\lambda \in C$ such that $a\lambda - c\lambda = \lambda$. Then define $\phi(\lambda, b) = (a\lambda - b, b - c\lambda)$. To see that this is an injection we show how to reconstruct λ and b given $a_\lambda - b$ and $b - c_\lambda$: First we have $\lambda = (a_\lambda - b)+(b-c_\lambda)$, so we obtain a_λ, c_λ, and thus b.

We now use $|A - C|\ |B| \le |A - B|\ |B - C|$, to obtain all sorts of useful inequalities:

- Taking $C = A$ gives $|A-A| \le |A-B|\ 2/|B|$.
- Then taking B = −A $\frac{|A-A|}{|A|} \le \left(\frac{|A+A|}{|A|}\right)^2$, which is the lower bound in, $\frac{1}{2} \le \log\left(\frac{|A+A|}{|A|}\right)/\log\left(\frac{|A-A|}{|A|}\right) \le 3.$
- Next taking $A = rA, B = -A$ with $C = -sA$ and then $C = sA$ implies:

$$|(r + s)A|\ |-A| \le |(r + 1)A|\ |sA - A|,$$
$$|rA - sA|\ |-A| \le |(r + 1)A|\ |(s + 1)A|.$$

With the choices $r = n-2$, $s = 2$, and $r = 2, s = 1$, respectively, we obtain,

$$\frac{|nA|}{|A|} \le \frac{|(n-1)A|}{|A|}\frac{|2A-A|}{|A|} \le \frac{|(n-1)A|}{|A|}\frac{|3A|}{|A|}\frac{|2A|}{|A|}$$

We deduce that, for n ≥ 3,

$$\frac{|nA|}{|A|} \le \left(\frac{|3A|}{|A|}\right)^{n-2}\left(\frac{|2A|}{|A|}\right)^{n-3} \quad \text{for all } n \ge 3;$$

and then that $$\frac{|rA-sA|}{|A|} \le \left(\frac{|3A|}{|A|}\right)^{r+s-2}\left(\frac{|2A|}{|A|}\right)^{r+s-4} \quad \text{for all } r,s \ge 2$$

We wanted bounds as a function of r, s and $|2A|/|A|$, and instead we have very easily obtained bounds in terms of these variables and $|3A|/|A|$. So the question becomes whether one can find an easy way to bound $|3A|/|A|$ in terms of $|2A|/|A|$? Certainly such bounds can be proved by straightforward combinatorial arguments, but we know of no proof that is quite so simple as that above. (Taking $r=1$, $s=2$ in the inequalities above, we see that we could replace $3A$ by $2A - A$ in these last few comments).

Representation Numbers

Denote the number of representations of n as a sum $a + b$, $a \in A$, $b \in B$ by,

$$r_{A+B}(n) := \#\{(a, b) : a \in A, b \in B, n = a + b\},$$

and similarly $r_{kA+\ell B}(n)$, etc. There are several straightforward but useful identities: First, by counting all ordered pairs (a, b), $a \in A$, $b \in B$ we obtain,

$$|A||B| = \sum_x r_{A+B}(x) = \sum_y r_{A-B}(y).$$

The solutions to $a+b=a'+b'$ with $a, a' \in A$, $b, b' \in B$ are the same as the solutions to $a-b'=a'-b$, which are the same as the solutions to $a-a'=b'-b$, and so,

$$E(A, B) := \sum_x r_{A+B}(x)^2 = \sum_y r_{A-B}(y)^2 = \sum_z r_{A-A}(z) r_{B-B}(z).$$

Therefore we obtain, by the Cauchy-Schwarz inequality, that,

$$(|A||B|)^2 = \left(\sum_x r_{A\pm B}(x) \right)^2 \le |A \pm B| E(A, B).$$

Also note that,

$$E(A, B) \le \begin{cases} \tfrac{1}{2} \max_x r_{A+B}(x) \Sigma_x r_{A+B}(x) = |A||B| \max_x r_{A+B}(x), \\ |A + B| \max_x r_{A+B}(x)^2. \end{cases}$$

Now we show that,

$$r_{A+B}(x) \le \frac{|A - B|^2}{|A + B|}$$

by exhibiting, for a given value of $x \in A+B$, an injection from,

$$R_{A+B}(x) \times (A+B) \to (A-B) \times (A-B)$$

where $R_{A+B}(x)$ is the set of representations of x as $a+b$, $a \in A$, $b \in B$. So fix a representation $a + b = x$, and for any $\lambda \in A+B$ fix $a_\lambda \in A$, $b_\lambda \in B$ such that $a_\lambda + b_\lambda = \lambda$. The map $(a, b, \{a_\lambda, b_\lambda\}) \to (a - b_\lambda, a_\lambda - b)$ is, indeed, an injection, because we can reconstruct our pre-image by noting that $\lambda = x + (a_\lambda - b) - (a - b_\lambda)$, from which we obtain a_λ and b_λ, then $a = (a - b_\lambda) + b_\lambda$ and $b = x - a$.

Combining the last three displayed equations we obtain,

$$|A+B| \le \frac{|A-B|^2}{\max_x r_{A+B}(x)} \le \frac{|A-B|^2 |A||B|}{E(A,B)} \le \frac{|A-B|^3}{|A||B|}.$$

Taking B = A gives the upper bound in (1.2).

Disjoint Unions

Lemma: There exists $X \subset B$ with $|X| \le |A + B| / |A|$ such that $B \subset A - A + X$.

Proof: $X \subset B$ to be as large as possible so that the sets $\{A + x : x \in X\}$ are disjoint. The union of these sets contains exactly $|A||X|$ elements, all in A + B, which implies that $|A| \cdot |X| \le |A + B|$.

Now if $b \in B$ then $(A + b) \cap (A + x) \neq \varnothing$ for some $x \in X$, else X would not have been maximal, so $b \in A - A + x$, and we are done.

Take $B = A - 2A$ in Lemma to get $2A - A \subset A - A + X$ where $X \subset 2A - A$ with,

$$|X| \le |2A - 2A| / |A|$$

(replacing X by –X for convenience). Add A to both sides to get,

$$3A - A \subset 2A - A + X \subset A - A + 2X$$

and then, proceeding by induction, we obtain,

$$mA - nA \subset A - A + (m - 1)X - (n - 1)X \text{ for all } m, n \ge 1.$$

Now, since each $|rX| \le |X|^r$, and as $|X| \le \frac{|2A - 2A|}{|A|}$, we deduce that,

$$\frac{|mA - nA|}{|A|} \le \frac{A - A|}{|A|} \left(\frac{2A - 2A}{|A|} \right)^{m+n-2} \quad \text{for all } m, n \ge 1$$

Another argument based on something similar to, but more complicated than, the above lemma, leads to the inequality,

$$|2B - 2B| \le |A + B|^4 \, |A - A| / |A|^4$$

Taking B = A in this formula, and then the first inequality in,

$$\frac{1}{2} \le \log\left(\frac{|A+A|}{|A|}\right) / \log\left(\frac{|A-A|}{|A|}\right) \le 3$$

we deduce from $\frac{|mA - nA|}{|A|} \le \frac{A - A|}{|A|}\left(\frac{2A-2A}{|A|}\right)^{m+n-2}$ for all m, n ≥ 1, that,

$$\frac{|mA-nA|}{|A|} \le \left(\frac{|2A|}{|A|}\right)^{6m+6m-10} \quad \text{for all } m,n \ge 1$$

Finally, selecting $A=(n-1)A, C=-A, B=A-A$ in $|A - C|\ |B| \le |A - B|\ |B - C|$, and then substituting in $\frac{|mA - nA|}{|A|} \le \frac{A - A|}{|A|}\left(\frac{2A-2A}{|A|}\right)^{m+n-2}$ for all m, n ≥ 1, we obtain,

$$\frac{|nA|}{|A|} \le \frac{|A-A|}{|A|}\left(\frac{|2A-2A|}{|A|}\right)^{n} \le \left(\frac{|2A|}{|A|}\right)^{6n+2} \quad \text{for all } n \ge 1$$

The strongest version of such an inequality that is known was first proved by Plünnecke [Pl], whose proof has been streamlined, over the years, by Ruzsa [R1] and others.

The Plünnecke-Ruzsa Theorem

For any m, n ≥ 0 we have,

$$\frac{|mA-nA|}{|A|} \le \left(\frac{|2A|}{|A|}\right)^{m+n}$$

We may rephrase this as: If $|2A| \le C|A|$ then $|mA - nA| \le C^{m+n}|A|$.

This result can be given in the slightly stronger form: If $|A + B| \le C|A|$ then $|mB - nB| < C^{m+n}|A|$ for all m, n ≥ 0. Taking B = A gives the above result. Taking B = −A implies that the assumption $|A - A| \le C|A|$ yields the same conclusion, and therefore we may replace the "≤ 3" by "≤ 2" in $\frac{1}{2} \le \log\left(\frac{|A+A|}{|A|}\right) / \log\left(\frac{|A-A|}{|A|}\right) \le 3$.

The Freiman-Ruzsa Theorem in Groups and where the Elements have Bounded Order

Take the union of,

$$mA - nA \subset A - A + (m-1)X - (n-1)X \text{ for all } m, n \ge 1$$

over all m, n ≥ 1 to obtain $\langle A\rangle \subset A - A + \langle X\rangle$. However $X \subset 2A - A \subset \langle A\rangle$ and so,

$$\langle A\rangle = A - A + \langle X\rangle.$$

Suppose that $|2A| \le C|A|$. Then $|X| \le |2A-2A|/|A| \le C4$ by the Plünnecke -Ruzsa theorem (we can get $\le C^6$ if we only use the results that are proved above). That is, the GAP $\langle A\rangle$ belongs to a union of translates of the GAP $\langle X\rangle$, which has (bounded) dimension

$\leq C^4$. If $A \subset G$, an abelian group in which the maximal order of any element is $\leq r$, then $|\langle X \rangle| \leq r\,|X|$. Therefore,

$$|\langle A \rangle| \leq |A - A|\,|\langle X \rangle| \leq C^2 |A| r^{|X|} \leq \left(C^2 r C^4\right)|A|.$$

The Balog-Szemeredi-(Gowers) Theorem

In many applications one does not have that A + B is small, but rather that there is a large subset $G \subset \{(a, b) : a \in A, b \in B\}$ which contains $\geq |A||B|$ elements, for which $SG := \{a+b : (a,b) \in G\}$ is small. One then wishes to conclude something about the structure of large subsets of A and B. In the case that $|A| = |B|$ there is an important result of Balog and Szemeredi [BS], strengthened by Gowers [G1] (and subsequently by several others) with a much easier proof – Here we simply state a version of this very flexible result, in order to get the flavour: Suppose that $|A| = |B| = n$ and that there exists $G \subset \{(a, b) : a \in A, b \in B\}$ containing $\geq \alpha n^2$ elements, for which $SG := \{a+b : (a, b) \in G\} \leq n$. Then there exists $A' \subset A$, $B' \subset B$ with $|A'|, |B'| \geq (\alpha/16)n$ for which $|A' + B'| \leq (2^{23}/\alpha^5)n$, with,

$$G \cap \{(a', b') : a o \in A', b' \in B'\}| \geq (\alpha\, 2\, /128)n^2$$

Discrete Fourier Transforms

One of the most useful tools in additive combinatorics are Fourier transforms in $\mathbb{Z}/N\mathbb{Z}$: For a function f: $\mathbb{Z}/N\mathbb{Z} \to \mathbb{C}$ we define,

$$\hat{f}(r) = \frac{1}{N}\sum_{s=0}^{N-1} f(s) e\left(\frac{rs}{N}\right)$$

where $e(t) = \exp(2i\pi t)$. This has inverse,

$$f(s) = \frac{1}{N}\sum_{r=0}^{N-1} \hat{f}(r) e\left(\frac{-rs}{N}\right)$$

One has,

$$\sum_r \hat{f}(r)\bar{\hat{g}}(r) = N\sum_r f(r)\bar{g}(r)$$

Parseval's identity is the case f = g, namely $\Sigma_r |\hat{f}(r)|^2 - N\Sigma_r |f(r)|^2$.

We define the convolution of two functions to be,

$$(f * g)(r) = \sum_{t-u=r} f(t)\overline{g(u)}$$

so that $\widehat{(f*g)} = \hat{f}\bar{\hat{g}}$, and,

$$N\sum_{r}|(f*g)(r)|^{2}=\sum_{r}|\hat{f}(r)|^{2}|\hat{g}('r)|^{2}$$

Taking g = f we obtain,

$$\sum_{r}|\hat{f}(r)|^{4}=N\sum_{a+b=c+d}f(a)f(b)\overline{f(c)f(d)}.$$

Let A be a subset of $\mathbb{Z}/N\mathbb{Z}$, and then define A(n) to be the characteristic function of A; that is, $A(n)=1$ if $n\in A$, and $A(n)=0$ otherwise. Hence,

$$\hat{A}(m)=\sum_{a\in A}e\left(\frac{am}{N}\right)$$

Noting that $(A*B)(n) = r_{A-B}(n)$ we deduce that,

$$E(A,B)=\sum_{n}r_{A-B}(n)^{2}=\sum_{n}|(A*B)(n)|^{2}\frac{1}{N}\sum_{n}|\hat{A}(n)|^{2}\,\hat{B}(n)|^{2}.$$

We also have,

$$\hat{A}(m)\hat{B}(m)=\sum_{n}r_{A+B}(n)e\left(\frac{mn}{N}\right),$$

which can be inverted to give,

$$r_{A+B}(n)=\frac{1}{N}\sum_{m}\hat{A}(m)\hat{B}(m)e\left(\frac{-mn}{N}\right);$$

a special case of which is,

$$r_{\kappa A-\kappa A}(n)=\frac{1}{N}\sum_{m}|\hat{A}(m)|^{2\kappa}\,e\left(\frac{-mn}{N}\right).$$

Basic Notions

Operations on Sets

Let A and B be finite subsets of an abelian group, then the sum set is defined to be,

$$A+B=\{a+b:a\in A,b\in B\}.$$

For example, we can write $\{1,2,3,4\}+\{1,2,3\}=\{2,3,4,5,6,7\}$. Similarly we can define the difference set of A and B to be,

$$A-B=\{a-b:a\in A,b\in B\}.$$

Here we provide other useful notations.

$$kA = \underbrace{A + A + \cdots + A}_{k \text{ terms}}$$

$$k \cdot A = \{ka : a \in A\}$$

Doubling Constant

Let A be a subset of an abelian group. The doubling constant measures how big the sum set |A + A| is compared to its original size |A|. We define the doubling constant of A to be,

$$K = \frac{|A + A|}{|A|}.$$

Ruzsa Distance

Let A and B be two subsets of an abelian group. We define the Ruzsa distance between these two sets to be the quantity,

$$d(A,B) = \log \frac{|A - B|}{\sqrt{|A||B|}}.$$

Ruzsa triangle inequality tells us that the Ruzsa distance obeys the triangle inequality:

$$d(B,C) \le d(A,B) + d(A,C).$$

However, since d(A, A) cannot be zero, note that the Ruzsa distance is not actually a metric.

RUZSA TRIANGLE INEQUALITY

In additive combinatorics, the Ruzsa triangle inequality, also known as the Ruzsa difference triangle inequality to differentiate it from some of its variants, bounds the size of the difference of two sets in terms of the sizes of both their differences with a third set. It was proven by Imre Ruzsa, and is so named for its resemblance to the triangle inequality. It is an important lemma in the proof of the Plünnecke-Ruzsa inequality.

If A and B are subsets of an abelian group, then the sumset notation A +B is used to denote $\{a + b : a \in A, b \in B\}$. Similarly, A – B denotes $\{a - b : a \in A, b \in B\}$.

Theorem (Ruzsa triangle inequality): If A, B, and C are finite subsets of an abelian group, then,

$$|A||B-C| \leq |A-B||A-C|.$$

An alternate formulation involves the notion of the Ruzsa distance.

If A and B are finite subsets of an abelian group, then the Ruzsa distance between these two sets, denoted d(A, B), is defined to be,

$$d(A,B) = \log\frac{|A-B|}{\sqrt{|A||B|}}.$$

Then, the Ruzsa triangle inequality has the following equivalent formulation:

Theorem (Ruzsa triangle inequality): If A, B, and C are finite subsets of an abelian group, then,

$$d(B,C) \leq d(A,B) + d(A,C).$$

This formulation resembles the triangle inequality for a metric space; however, the Ruzsa distance does not define a metric space since d(A, A) is not always zero.

Proof: To prove the statement, it suffices to construct an injection from the set $A\times(B-C)$ to the set $(A-B)\times(A-C)$. Define a function ϕ as follows. For each $x \in B-C$ choose a $b(x) \in B$ and a $c(x) \in C$ such that $x = b(x)-c(x)$. By the definition of B – C, this can always be done. Let $\phi: A\times(B-C) \rightarrow (A-B)\times(A-C)$ be the function that sends (a, x) to $(a-b(x), a-c(x))$. For every point $\phi(a,x) = (y,z)$ in the set is $(A-B)\times(A-C)$, it must be the case that $x = z-y$ and $a = y+b(x)$. Hence, ϕ maps every point in $(A-B)\times(A-C)$ to a distinct point in $(A-B)\times(A-C)$ and is thus an injection. In particular, there must be at least as many points in $(A-B)\times(A-C)$ as in $A\times(B-C)$. Therefore,

$$|A||B-C| = |A\times(B-C)| \leq |(A-B)\times(A-C)| = |A-B||A-C|,$$

completing the proof.

Variants of the Ruzsa Triangle Inequality

The Ruzsa sum triangle inequality is a corollary of the Plünnecke-Ruzsa inequality (which is in turn proved using the ordinary Ruzsa triangle inequality).

Theorem (Ruzsa sum triangle inequality): If A, B< and C are finite subsets of an abelian group, then,

$$|A||B+C| \leq |A+B||A+C|.$$

Proof: The proof uses the following lemma from the proof of the Plünnecke-Ruzsa inequality.

Lemma: Let A and B be finite subsets of an abelian group G. If $X \subseteq A$ is a nonempty subset that minimizes the value of $K' = |X+B|/|X|$, then for all finite subsets $C \subset G$,

$$|X+B+C| \leq K'|X+C|.$$

If A is the empty set, then the left side of the inequality becomes 0, so the inequality is true. Otherwise, let X be a subset of A that minimizes $K' = |X+B|/|X|$. Let $K = |A+B|/|A|$. The definition of X implies that $K' \leq K$. Because $X \subset A$, applying the above lemma gives:

$$|B+C| \leq |X+B+C| \leq K'|X+C| \leq K'|A+C| \leq K|A+C| = \frac{|A+B||A+C|}{|A|}.$$

Re-arranging gives the Ruzsa sum triangle inequality.

By replacing B and C in the Ruzsa triangle inequality and the Ruzsa sum triangle inequality with $-B$ and $-C$ as needed, a more general result can be obtained: If A, B, and C are finite subsets of an abelian group then,

$$|A||B \pm C| \leq |A \pm B||A \pm C|,$$

where all eight possible configurations of signs hold. These results are also sometimes known collectively as the Ruzsa triangle inequalities.

GOWERS NORM

In mathematics, in the field of additive combinatorics, a Gowers norm or uniformity norm is a class of norms on functions on a finite group or group-like object which quantify the amount of structure present, or conversely, the amount of randomness. They are used in the study of arithmetic progressions in the group. It is named after Timothy Gowers, who introduced it in his work on Szemerédi's theorem.

Let f be a complex-valued function on a finite Abelian group G and let J denote complex conjugation. The Gowers d-norm is,

$$\| f \|_{U^d(G)}^{2^d} = \mathbb{E}_{x,h_1,\dots,h_d \in G} \prod_{\omega_1,\dots,\omega_d \in \{0,1\}} J^{\omega_1+\dots+\omega_d} f\left(x + h_1\omega_1 + \dots + h_d\omega_d\right).$$

Gowers norms are also defined for complex valued functions f on a segment $[N]=\{0,1,2,\dots,N-1\}$, where N is a positive integer. In this context, the uniformity norm is given as $\backslash f \backslash_{U^d[N]} = \backslash \tilde{f} \backslash_{U^d(\mathbb{Z}/\tilde{N}\mathbb{Z})} / \backslash 1_{[N]} \backslash_{U^d(\mathbb{Z}/\tilde{N}\mathbb{Z})}$, where $\tilde{N}$ is a large integer, $1_{[N]}$ denotes the indicator function of [N], and $f(x)$ is equal to $f(x)$ for $x \in [N]$ and 0 for all other x. This definition does not depend on $\tilde{N}$, as long as $\tilde{N} > 2^d N$.

Inverse Conjectures

An inverse conjecture for these norms is a statement asserting that if a bounded function f has a large Gowers d-norm then f correlates with a polynomial phase of degree d-1 or other object with polynomial behaviour (e.g. a (d-1)-step nilsequence). The precise statement depends on the Gowers norm under consideration.

The Inverse Conjecture for vector spaces over a finite field $\delta > 0$ asserts that for any $c > 0$ there exists a constant $\mathbb{F}$ such that for any finite dimensional vector space V over $\mathbb{F}$ and any complex valued function f on V, bounded by 1, such that $\backslash f \backslash_{U^d[V]} \geq \delta$, there exists a polynomial sequence $P: V \to \mathbb{R}/\mathbb{Z}$ such that,

$$\left|\frac{1}{|V|}\sum_{x \in V} f(x)e(-P(x))\right| \geq c,$$

where $e(x) := e^{2\pi i x}$. This conjecture was proved to be true by Bergelson, Tao, and Ziegler.

The Inverse Conjecture for Gowers $U^d[N]$ norm asserts that for any $\delta > 0$, a finite collection of (d-1)-step nilmanifolds $\mathcal{M}_\delta$ and constants c, C can be found, so that the following is true. If N is a positive integer and $f:[N] \to \mathbb{C}$ is bounded in absolute value by 1 and $\|f\|_{U^d[N]} \geq \delta$, then there exists a nilmanifold $G/\Gamma \in \mathcal{M}_\delta$ and a nilsequence $F(g^n x)$ where $g \in G$, $x \in G/\Gamma$ and $F: G/\Gamma \to \mathbb{C}$ bounded by 1 in absolute value and with Lipschitz constant bounded by C such that:

$$\left|\frac{1}{N}\sum_{n=0}^{N-1} f(n)\overline{F(g^n x)}\right| \geq c.$$

This conjecture was proved to be true by Green, Tao, and Ziegler. It should be stressed that the appearance of nilsequences in the above statement is necessary. The statement is no longer true if we only consider polynomial phases.

PLÜNNECKE–RUZSA INEQUALITY

In additive combinatorics, the Plünnecke-Ruzsa inequality is an inequality that bounds the size of various sumsets of a set B, given that there is another set A so that A + B is not much larger than A. A slightly weaker version of this inequality was originally proven and published by Helmut Plünnecke. Imre Ruzsa later published a simpler proof of the current, more general, version of the inequality. The inequality forms a crucial step in the proof of Freiman's theorem.

The following sumset notation is standard in additive combinatorics. For subsets A and B of an abelian group and a natural number k, the following are defined:

$$A + B = \{a + b : a \in A, b \in B\}$$

$$A - B = \{a - b : a \in A, b \in B\}$$

$$kA = \underbrace{A + A + \cdots + A}_{k \text{ times}}$$

The set A + B is known as the sumset of A and B.

The most commonly cited version of the statement of the Plünnecke-Ruzsa inequality is the following:

Theorem (Plünnecke-Ruzsa inequality): If A and B are finite subsets of an abelian group and K is a constant so that $|A+B| \leq K|A|$, then for all nonnegative integers m and n,

$$|mB - nB| \leq K^{m+n}|A|.$$

This is often used when A = B, in which case the constant K = |2A|/|A| is known as the doubling constant of A. In this case, the Plünnecke-Ruzsa inequality states that sumsets formed from a set with small doubling constant must also be small.

Plünnecke's Inequality

The version of this inequality that was originally proven by Plünnecke is slightly weaker.

Theorem (Plünnecke's inequality): Suppose A and B are finite subsets of an abelian group and K is a constant so that $|A+B| \leq K|A|$. Then for all nonnegative integer m. $|mB| \leq K^m|A|$.

Proof of Ruzsa Triangle Inequality

The Ruzsa triangle inequality is an important tool which is used to generalize Plünnecke's inequality to the Plünnecke-Ruzsa inequality. Its statement is:

Theorem (Ruzsa triangle inequality): If A, B, and C are finite subsets of an abelian group, then,

$$|A||B-C| \leq |A-B||A-C|.$$

Proof of Plünnecke-Ruzsa Inequality

The following simple proof of the Plünnecke-Ruzsa inequality is due to Petridis:

Lemma: Let A and B be finite subsets of an abelian group G. If $X \subseteq A$ is a nonempty subset that minimizes the value of $K' = |X+B|/|X|$, then for all finite subsets $C \subset G$,

$$|X+B+C| \leq K'|X+C|.$$

Proof: This is demonstrated by induction on the size of $|C|$. For the base case of $|C|=1$, note that S + C is simply a translation of S for any $S \subseteq G$, so,

$$|X+B+C|=|X+B|=K'|X|=K'|X+C|.$$

For the inductive step, assume the inequality holds for all $C \subseteq G$ with $|C| \leq n$ for some positive integer n. Let C be a subset of G with $|C|=n+1$, and let $C=C' \sqcup \{\gamma\}$ for some $\gamma \in C$. (In particular, the inequality holds for C'). Finally, let,

$$Z=\{x \in X : x+B+\{\gamma\} \subseteq X+B+C'\}.$$

The definition of Z implies that $Z+B+\{\gamma\} \subseteq X+B+C'$. Thus, by the definition of these sets,

$$X+B+C=(X+B+C') \cup ((X+B+\{\gamma\}) \setminus (Z+B+\{\gamma\})).$$

Hence, considering the sizes of the sets,

$$\begin{aligned}|X+B+C| &\leq |X+B+C'|+|(X+B+\{\gamma\}) \setminus (Z+B+\{\gamma\})| \\ &= |X+B+C'|+|X+B+\{\gamma\}|-|Z+B+\{\gamma\}| \\ &= |X+B+C'|+|X+B|-|Z+B|.\end{aligned}$$

The definition of Z implies that $Z \subseteq X \subseteq A$, so by the definition of X, $|Z+B| \geq K'|Z|$. Thus, applying the inductive hypothesis on C' and using the definition of X,

$$\begin{aligned}|X+B+C| &\leq |X+B+C'|+|X+B|-|Z+B| \\ &\leq K'|X+C'|+|X+B|-|Z+B| \\ &\leq K'|X+C'|+K'|X|-|Z+B| \\ &\leq K'|X+C'|+K'|X|-K'|Z| \\ &= K'(|X+C'|+|X|-|Z|).\end{aligned}$$

To bound the right side of this inequality, let $W=\{x \in X : x+\gamma \in X+C'\}$. Suppose $y \in X+C'$ and $y \in X+\{\gamma\}$, then there exists $x \in X$ such that $x+\gamma=y \in X+C'$. Thus, by definition, $x \in W$, so $y \in W+\{\gamma\}$. Hence, the sets $X+C'$ and $(X+\{\gamma\}) \setminus (W+\{\gamma\})$ are disjoint. The definitions of W and C' thus imply that,

$$X+C=(X+C') \sqcup ((X+\{\gamma\}) \setminus (W+\{\gamma\})).$$

Again by definition, $W \subseteq Z$, so $|W| \leq |Z|$. Hence,

$$\begin{aligned}|X+C| &= |X+C'|+|(X+\{\gamma\}) \setminus (W+\{\gamma\})| \\ &= |X+C'|+|X+\{\gamma\}|-|W+\{\gamma\}| \\ &= |X+C'|+|X|-|W| \\ &\geq |X+C'|+|X|-|Z|.\end{aligned}$$

Putting the above two inequalities together gives,

$$|X+B+C| \le K'(|X+C'|+|X|-|Z|) \le K'|X+C|.$$

This completes the proof of the lemma.

To prove the Plünnecke-Ruzsa inequality, take X and K' as in the statement of the lemma. It is first necessary to show that,

$$|X+nB| \le K^n |X|.$$

This can be proved by induction. For the base case, the definitions of K and K' imply that $K' \le K$. Thus, the definition of X implies that $|X+B| \le K|X|$. For inductive step, suppose this is true for $n=j$. Applying the lemma with $C=jB$ and the inductive hypothesis gives,

$$|X+(j+1)B| \le K'|X+jB| \le K|X+jB| \le K^{j+1}|X|.$$

This completes the induction. Finally, the Ruzsa triangle inequality gives,

$$|mB-nB| \le \frac{|X+mB||X+nB|}{|X|} \le \frac{K^m|X|K^n|X|}{|X|} = K^{m+n}|X|.$$

Because $X \subseteq A$, it must be the case that $|X| \le |A|$. Therefore,

$$|mB-nB| \le K^{m+n}|A|.$$

This completes the proof of the Plünnecke-Ruzsa inequality.

Plünnecke Graphs

Both Plünnecke's proof of Plünnecke's inequality and Ruzsa's original proof of the Plünnecke-Ruzsa inequality use the method of Plünnecke graphs. Plünnecke graphs are a way to capture the additive structure of the sets $A, A+B, A+2B, \ldots$ in a graph theoretic manner First important to defining Plünnecke graphs is the notion of a commutative graph.

A directed graph G is called semicommutative if, whenever there exist distinct $x, y, z_1, z_2, \ldots, z_k$ such that (x,y) and (y,z_i) are edges in G, for each i, then there also exist distinct $y_1, y_2, \ldots, y_k$ so that (x,y_i) and (y_i,z_i) are edges in G for each i. G is called commutative if it is semicommutative and the graph formed by reversing all its edges is also semicommutative.

A layered graph is a (directed) graph G whose vertex set can be partitioned,

$$V_0 \cup V_1 \cup \ldots \cup V_m$$

so that all edges in G are from V_i to V_{i+1} for some i. A Plünnecke graph is a layered graph which is commutative.

The relevant example of a Plünnecke graph is the following, showing how the structure of the sets $A, A+B, A+2B, \ldots, A+mB$ is a case of that of a Plünnecke graph.

Example: Let A, B be subsets of an abelian group. Then, let G be the layered graph so that each layer V_j is a copy of $A+jB$, so that $V_0 = A$, $V_1 = A+B$,..., $V_m = A+mB$. Create the edge (x,y) (where $x \in V_i$ and $y \in V_{i+1}$) whenever there exists $b \in B$ such that $y = x+b$. (In particular, if $x \in V_i$, then $x+b \in V_{i+1}$ by definition, so every vertex has out-degree equal to the size of B). Then G is a Plünnecke graph. For example, to check that G is semicommutative, if (x,y) and (y,z_i) are edges in G for each i, then $y-x, z_i - y \in B$. Then, let $y_i = x+z_i - y$, so that $y_i - x = z_i - y \in B$ and $z_i - y_i = y - x \in B$. Thus, G is semi-commutative. It can be similarly checked that the graph formed by reversing all edges of G is also semicommutative, so G is a Plünnecke graph.

In a Plünnecke graph, the image of a set $X \subseteq V_0$ in V_j, written $\mathrm{im}(X, V_j)$, is defined to be the set of vertices in V_j which can be reached by a path starting from some vertex in X. In particular, in the aforementioned example, $\mathrm{im}(X, V_j)$ is just $X+jB$.

The magnification ratio between V_0 and V_j, denoted $\mu_j(G)$, is then defined as the minimum factor by which the image of a set must exceed the size of the original set. Formally,

$$\mu_j(G) = \min_{X \subseteq V_0, X \neq \varnothing} \frac{|\mathrm{im}(X, V_j)|}{|X|}.$$

Plünnecke's theorem is the following statement about Plünnecke graphs.

Theorem (Plünnecke›s theorem): Let G be a Plünnecke graph. Then, $\mu_j(G)^{1/j}$ is decreasing in j.

The proof of Plünnecke's theorem involves a technique known as the "tensor product trick", in addition to an application of Menger's theorem.

The Plünnecke-Ruzsa inequality is a fairly direct consequence of Plünnecke's theorem and the Ruzsa triangle inequality. Applying Plünnecke's theorem to the graph given in the example, at $j = m$ and $j = 1$, yields that if $|A+B|/|A| = K$, then there exists $X \subseteq A$ so that $|X+mB|/|X| \leq K^m$. Applying this result once again with X instead of A, there exists $X' \subseteq X$ so that $|X'+nB|/|X'| \leq K^m$. Then, by Ruzsa's triangle inequality (on $-X', mB, nB$),

$$|mB - nB| \leq |X' + mB||X' + nB||X'| \leq K^{m+n}|X| \leq K^{m+n}|X|,$$

thus proving the Plünnecke-Ruzsa inequality.

SUM-FREE SEQUENCE

In mathematics, a sum-free sequence is an increasing positive integer sequence, $\{n_k\}_{k \in \mathbb{N}}$ such that for each k, n_k cannot be represented as a sum of any subset of the preceding elements of the same sequence.

The definition of sum-free sequence is different of that of sum-free set, because in a sum-free set only the sums of two elements must be avoided, while a sum-free sequence must avoid sums of larger sets of elements as well.

Example: The powers of two, 1, 2, 4, 8, 16,... form a sum-free sequence: each term in the sequence is one more than the sum of all preceding terms, and so cannot be represented as a sum of preceding terms.

Sums of Reciprocals

A set of integers is said to be small if the sum of its reciprocals converges to a finite value. For instance, by the prime number theorem, the prime numbers are not small. Paul Erdős proved that every sum-free sequence is small, and asked how large the sum of reciprocals could be. For instance, the sum of the reciprocals of the powers of two (a geometric series) is two.

If R denotes the maximum sum of reciprocals of a sum-free sequence, then through subsequent research it is known that $2.0654 < R < 2.8570$.

Density

It follows from the fact that sum-free sequences are small that they have zero Schnirelmann density; that is, if $A(x)$ is defined to be the number of sequence elements that are less than or equal to x, then $A(x) = o(x)$. Erdős showed that for every sum-free sequence there exists an unbounded sequence of numbers x_i for which $A(x_i) = O(x^{\varphi-1})$ where φ is the golden ratio, and he exhibited a sum-free sequence for which, for all values of x, $A(x) = \Omega(x^{2/7})$, subsequently improved to $A(x) = \Omega(x^{1/3})$ by Deshouillers, Erdős and Melfi in 1999 and to $A(x) = \Omega(x^{1/2-\varepsilon})$ by Luczak and Schoen, who also proved that the exponent 1/2 cannot be furthermore improved.

SUM-FREE SET

In additive combinatorics and number theory, a subset A of an abelian group G is said to be sum-free if the sumset A⊕A is disjoint from A. In other words, A is sum-free if the equation $a + b = c$ has no solution with $a, b, c \in A$.

For example, the set of odd numbers is a sum-free subset of the integers, and the set {N+1,..., 2N} forms a large sum-free subset of the set {1,...,2N}. Fermat's Last Theorem is the statement that, for a given integer n > 2, the set of all nonzero n^{th} powers of the integers is a sum-free subset.

Some basic questions that have been asked about sum-free sets are:

- How many sum-free subsets of {1,..., N} are there, for an integer N? Ben Green has shown that the answer is $O(2^{N/2})$, as predicted by the Cameron–Erdős conjecture.
- How many sum-free sets does an abelian group G contain?
- What is the size of the largest sum-free set that an abelian group G contains?

A sum-free set is said to be maximal if it is not a proper subset of another sum-free set.

RESTRICTED SUMSET

In additive number theory and combinatorics, a restricted sumset has the form. Let A, B, S be finite subsets of an abelian group G. Suppose that the restricted sumset,

$$C = \{a + b:\ a \in A,\ b \in B, \text{ and } a - b \notin S\}$$

is nonempty and some c ∈ C can be written as a + b with a ∈ A and b ∈ B in at most m ways. We show that if G is torsion-free or elementary abelian then $|C| \geq |A| + |B| - |S| - m$. We also prove that $|C| \geq |A| + |B| - 2|S| - m$ if the torsion subgroup of G is cyclic. In the case S = {0} this provides an advance on a conjecture of Lev.

Let A and B be finite nonempty subsets of an (additively written) abelian group G. The sumset of A and B is defined by,

$$A + B = \{a + b:\ a \in A \text{ and } b \in B\}.$$

The Cauchy-Davenport theorem, a basic result in additive combinatorial number theory, states that,

$$|A + B| \geq \min\{p, |A| + |B| - 1\}$$

if $G = \mathbb{Z}/p\mathbb{Z}$ with p prime. Another theorem due to Kemperman and Scherk asserts that,

$$|A + B| \geq |A| + |B| - \min_{c \in A+B} \nu_{A,B}(c),$$

where,

$$\nu_{A,B}(c) = \left|\{(a,\ b) \in A \times B:\ a+b=c\}\right|;$$

in particular, we have $|A+B| \geq |A|+|B|-1$ if some c ∈ A + B can be uniquely written as a + b with a ∈ A and b ∈ B.

Now we define the restricted sumset,

$$A \dot{+} B = \{a+b:\ a \in A,\ b \in B,\ \text{and } a \neq b\}.$$

In 1964, Erdös and Heilbronn [EH] conjectured that if $G = \mathbb{Z}/p\mathbb{Z}$ with p prime then,

$$|A \dot{+} A| \geq \min\{p,\ 2|A|-3\}.$$

This is much more difficult than the Cauchy-Davenport theorem concerning unrestricted sumsets. It had been open for thirty years until Dias da Silva and Hamidoune [DH] confirmed it in 1994 using representations of symmetric groups. Later Alon, Nathanson and Ruzsa [ANR1, ANR2] developed a powerful polynomial method to give a simpler proof of the Erdös-Heilbronn conjecture. They showed that if $G = \mathbb{Z}/p\mathbb{Z}$ with p prime then,

$$|A \dot{+} B| \geq \min\{p, |A|+|B|-2-\delta\},$$

where δ is 1 or 0 according to whether $|A|=|B|$ or not. The reader may consult [HS], [K1], [K2], [L1], [LS] and [SY] for various extensions of the Erdös-Heilbronn conjecture.

Motivated by the Kemperman-Scherk theorem and the Erd″os-Heilbronn conjecture, Lev [L2] proposed the following interesting conjecture.

Conjecture: (Lev). Let G be an abelian group, and let A and B be finite nonempty subsets of G. Then we have,

$$|A \dot{+} B| > |A|+|B|-2-\min_{c \in A+B} \nu_{A,B}(c).$$

This conjecture is known to be true for torsion-free abelian groups and elementary abelian 2-groups. It also holds when $|G|$ is prime, or G is cyclic and $|G| \leq 25$.

Theorem: Let A and B be finite nonempty subsets of a field F. Let P(x, y) ∈ F[x, y] and,

$$C = \{a+b:\ a \in A,\ b \in B,\ \text{and } P(a,\ b) \neq 0\}$$

If C is nonempty, then,

$$C| \geq |A|+|B|-\deg P-\min_{c \in C} \nu_{A,B}(c).$$

When $P(x, y)=1$, and equation above $C| \geq |A|+|B - \deg P - \min_{c \in C} \nu_{A,B}(c)$ becomes equation $A+B| \geq |A|+|B| - \min_{c \in A+B} \nu_{A,B}(c)$.

The difference between the minima in equation $|A \dot{+} B| > |A|+|B|-2-\min_{c \in A+B} \nu_{A,B}(c)$ and $C| \geq |A|+|B|-\deg P-\min_{c \in C} \nu_{A,B}(c)$: As $C \subseteq A + B$ we have,

$$\min_{c \in A+B} \nu_{A,B}(c) \leq \min_{c \in C} \nu_{A,B}(c).$$

Theorem: Let A and B be finite nonempty subsets of an abelian group G whose torsion subgroup,

$$\mathrm{Tor}(G)=\{g \in G : g \text{ has a finite order}\}$$

is cyclic. For $i = 1,\ldots, l$ let m_i and n_i be nonnegative integers and let $d_i \in G$. Suppose that,

$$C=\{a+b:\ a \in A,\ b \in B, \text{ and } m_i a - n_{ib} \neq d_i \text{ for all } i=1, \ldots, l\}$$

is nonempty. Then,

$$|C| \geq |A|+|B|-\sum_{i=1}^{l}(m_i+n_i)-\min_{c \in C} \nu_{A,B}(C).$$

When A and B are finite subsets of $\mathbb{Z}$, the restricted sumset in equation $C=\{a+b:\ a \in A,\ b \in B, \text{ and } m_i a - n_{ib} \neq d_i \text{ for all } i=1, \ldots, l\}$ was first studied by Sun.

From Theorems above we deduce the following result on difference restricted sumsets.

Theorem: Let G be an abelian group, and let A, B, S be finite nonempty subsets of G with,

$$C=\{a+b:\ a \in A,\ b \in B, \text{ and } a-b \notin S\} \neq \phi.$$

(i) If G is torsion-free or elementary abelian, then,

$$|C| \geq |A|+|B|-|S|-\min_{c \in C} \nu_{A,B}(c).$$

(ii) If Tor(G) is cyclic, then,

$$|C| \geq |A|+|B|-2|S|-\min_{c \in C} \nu_{A,B}(c).$$

Proof: Without loss of generality we can assume that G is generated by the finite set $A \cup B \cup S$.

If $G \cong \mathbb{Z}^n$, then we can simply view G as the ring of algebraic integers in an algebraic number field K with $[K: \mathbb{Q}] = n$. If $G \cong (\mathbb{Z}/p\mathbb{Z})^n$ n where p is a prime, then G is isomorphic to the additive group of the finite field with p^n elements. Thus part (i) follows from

Theorem above in the case $P(x, y) = \prod_{s \in S}(x - y - s)$. Let $d_1, \ldots, d_l$ be all the distinct elements of S. Applying previous theorem with $m_i = n_i = 1$ for all i = 1,..., l we immediately get the second part.

It is interesting to compare Theorem in the case S = {0} with previous conjecture.

Concerning the set C given by equation $C = \{a + b : a \in A, b \in B, \text{and } a - b \notin S\} \neq \phi$ there are some known results of different types. When A, B, S are finite nonempty subsets of a field whose characteristic is an odd prime p, the authors [PS] proved that $|C| \geq \min\{p, |A| + |B| - |S| - q - 1\}$, where q is the largest power of p not exceeding $|S|$. By modifying Károlyi's proof of previous theorem, we can show that if q > 1 is a power of a prime p, and A, B, S are subsets of $\mathbb{Z}/q\mathbb{Z}$ with min{|A|, |B|} > |S|, then $|C| \geq \min\{p, |A| + |B| - 2|S| - 1\}$.

Combinatorial Nullstellensatz: Let $A_1, \ldots, A_n$ be finite nonempty subsets of a field F, and set $g_i(x) = \prod_{a \in A_i}(x - a)$ for i = 1,..., n. Then $f(x_1, \ldots, x_n) \in F[x_1, \ldots, x_n]$ vanishes over the Cartesian product $A_1 \times \cdots \times A_n$ if and only if it can be written in the form:

$$f(x_1, \ldots, x_n) = \sum_{i=1}^{n} g_i(x_i) h_i(x_1, \ldots, x_n)$$

where $h_i(x_1, \ldots, x_n) \in F[x_1, \ldots, x_n]$ and $\deg h_i \leq \deg f - \deg g_i$.

With help of the Combinatorial Nullstellensatz, we provide a lemma for our purposes.

Lemma: Let A and B be finite nonempty subsets of a field F, and write:

$$v_i = |\{(a, b) \in A \times B : a + \lambda_i b = \mu_i\}|$$

for i = 1,..., k where $\lambda_i \in F \setminus \{0\}$ and $\mu_i \in F$. Let $P(x, y) \in F[x, y]$. Suppose that for any i = 1,..., k there are $a \in A$ and $b \in B$ with $P(a, b) \neq 0$ and $a + \lambda_i b = \mu_i$, and that for each $(a, b) \in A \times B$ with $P(a, b) \neq 0$ there is a unique $i \in \{1, \ldots, k\}$ with $a + \lambda_i b = \mu_i$. Then we have,

$$k + \min\{v_1, \ldots, v_k\} \geq |A| + |B| - \deg P.$$

Proof: Clearly,

$$f(x, y) := P(x, y) \prod_{j=1}^{k} (x + \lambda_j y - \mu_j)$$

vanishes over A × B. Set $g_A(x) = \prod_{a \in A}(x - a)$ and $g_B(y) = \prod_{b \in B}(y - b)$. By the Combinatorial Nullstellensatz, there are $h_A(x, y), h_B(x, y) \in F[x, y]$ such that,

$$f(x, y) = g_A(x) h_A(x, y) + g_B(y) h_B(x, y)$$

and,

$$\max\{\deg g_A + \deg h_A,\ \deg g_B + \deg h_B\} \le \deg f.$$

Fix $1 \le i \le k$. Write $h_B(x, y) = \sum_{s,t \ge 0} c^s tx^s y^t$ where $c_{st} \in F$. Then,

$$h_B(x, y) = \sum_{s,t\ge 0} cst\left((x+\lambda_i y-\mu_i)+\mu_i-\lambda_i y\right)^s y^t = (x+\lambda_i y-\mu_i)q(x, y)+r(y),$$

where $q(x, y) \in F[x, y]$, and $r(y) = h_B(\mu_i - \lambda_i y, y)$ has degree not greater than $\deg h_B$. Now assume that $k + v_i < |A| + |B| - \deg P$. We want to deduce a contradiction.

Set,

$$A_0 = \{a \in A : (\mu_i - a)/\lambda_i \notin B\}.$$

Obviously $|A_0| = |A| - v_i$ and $g_B((\mu_i - a)/\lambda_i) \ne 0$ for any $a \in A_0$. If $a \in A_0$, then,

$$g_B\left(\frac{\mu_i - a}{\lambda_i}\right) h_B\left(a, \frac{\mu_i - a}{\lambda_i}\right) = f\left(a, \frac{\mu_i - a}{\lambda_i}\right) - g_A(a)h_A\left(a, \frac{\mu_i - a}{\lambda_i}\right) = 0$$

and hence,

$$r\left(\frac{\mu_i - a}{\lambda_i}\right) = h_B\left(a, \frac{\mu_i - a}{\lambda_i}\right) = 0.$$

Since $\deg r \le \deg f - \deg g_B < |A| - v_i = |A_0|$,, we must have $r(y) = 0$, i.e., $h_B(x, y)$ is divisible by $x + \lambda_i y - \mu_i$. Recall that there are $a_0 \in A$ and $b_0 \in B$ such that $P(a_0, b_0) \ne 0$ and $a_0 + \lambda_i b_0 = \mu_i$. Since $h_B(a_0, b_0) = 0$, the polynomial,

$$P(a_0, y)\prod_{j=1}^{k}(a_0 + \lambda_j y - \mu_j) = f(a_0, y) = g_B(y)h_B(a_0, y)$$

is divisible by (y–bo) 2. As $a_0 + \lambda_j b_0 \ne \mu_j$ for any $j \ne i$, we must have $y - b_0 \mid P(a_0, y)$, which contradicts the fact that $P(a_0, b_0) \ne 0$.

Proofs of Theorems

Proof of Theorem. Let $\mu_1, \ldots, \mu_k$ be all the distinct elements of C. Applying Lemma with $\lambda_1 = \cdots = \lambda_k = 1$, we find that,

$$|C| + \min_{c\in C} v_{A,B}(c) \ge |A| + |B| - \deg P$$

which is equivalent to equation $C| \ge |A| + |B| - \deg P - \min_{c\in C} v_{A,B}(c)$.

Proof of previous Theorem: Without loss of generality, we can assume that G is finitely generated, and furthermore that G is a subgroup of the multiplicative group of the field of complex numbers; thus, C is the set,

$$\{ab : a \in A,\ b \in B, \text{ and } a^{m_i}\ b^{-n_i} \neq d_i \text{ for all } i = 1, \ldots, l\}.$$

Let $-\lambda_1, \ldots, -\lambda_k$ be all the distinct elements of C, and set,

$$P(x,y) = \prod_{i=1}^{l} \left(x^{m_i} y^{n_i} - d_i\right)$$

Then, for each $j \in \{1, \ldots, k\}$, there are $a \in A$ and $b \in B$ such that $a + \lambda j^{b-1} = 0$ and $P(a, b^{-1}) \neq 0$. If $a \in A$, $b \in B$ and $P(a, b^{-1}) \neq 0$, then there is a unique $j \in \{1, \ldots, k\}$ such that $\lambda_j = -ab$ (i.e., $a + \lambda_j b \ = 0$). Applying previous Lemma to the sets A and $B^{-1} = \{b^{-1}$: $b \in B\}$ with $\mu_1 = \cdots = \mu_k = 0$, we obtain that,

$$k + \min_{1 \le j \le k} \left|\{(a,\ b) \in A \times B:\ a + \lambda_j\ b^{-1} = 0\}\right| \geq |A| + |B^{-1}| - \deg P.$$

Therefore,

$$C| + \min_{c \in C} \left|\{(a,\ b) \in A \times B:\ ab = c\}\right| \geq |A| + |B| - \sum_{i=1}^{l} (m_i + n_i)$$

as desired.

References

- Tao, T.; Vu, V. (2006). Additive Combinatorics. Cambridge: Cambridge University Press. ISBN 978-0-521-85386-6
- Hartnett, Kevin. “Mathematicians Catch a Pattern by Figuring Out How to Avoid It”. Quanta Magazine. Retrieved 2019-11-26
- Restricted-sumsets-and-a-conjecture-of-Lev-2118744: researchgate.net, Retrieved 12 March, 2020
- Yang, Shi Chun (2009), “Note on the reciprocal sum of a sum-free sequence”, Journal of Mathematical Research and Exposition, 29 (4): 753–755, MR 2549677
- Green, Ben (November 2004). “The Cameron–Erdős conjecture”. Bulletin of the London Mathematical Society. 36 (6): 769–778. arXiv:math.NT/0304058. doi:10.1112/S0024609304003650. MR 2083752

Algebraic Combinatorics

CHAPTER 6

The area of mathematics that applies methods of abstract algebra, group theory and representation theory, to problems of combinatorics is called algebra combinatorics. Bender-Knuth involution, h-vector, Stanley's reciprocity theorem, Eulerian poset, Buekenhout geometry, etc. are some of its aspects. This chapter discusses algebraic combinatorics in detail.

Algebraic combinatorics use algebraic methods to solve combinatorial problems, or use combinatorial methods and ideas to study algebraic objects. The unifying feature of the subject is any significant interaction between algebraic and combinatorial ideas. As a simple example, to solve an enumeration problem one often encodes combinatorial data into an algebra of formal power series by means of a generating function. Algebraic manipulations with these power series then provide a systematic way to solve the original counting problem. Methods from complex analysis can be used to obtain asymptotic solutions even when exact answers are intractable.

As another example, group theory and linear algebra are used to understand the structure of graphs. Most graphs have no nontrivial symmetries (automorphisms) - graphs with many symmetries are highly structured and have applications in design theory, coding theory, and geometry. For any graph, the eigenvalues of its adjacency matrix encode a great deal of structural and enumerative information about it.

The examples above are applications of algebra to combinatorics. Conversely, the most concrete way to understand the ring of symmetric functions is through the combinatorics of Young tableaux. This ring can be used for many things: representation theory of the symmetric and general linear groups; intersection theory on Grassmann or flag manifolds; enumeration of maps (graphs) embedded on surfaces. Thus the combinatorics of Young tableaux (and related objects) describes complicated phenomena in representation theory, geometry, and enumeration.

These examples are far from exhaustive. There are many variations on the above themes, and applications of these ideas to statistical mechanics, high-energy physics, knot theory, algebraic geometry, probability theory, analysis of algorithms, and too many more areas.

COHERENT ALGEBRA

A coherent algebra is an algebra of complex square matrices that is closed under ordinary matrix multiplication, Schur product, transposition, and contains both the identity matrix I and the all-ones matrix J.

A subspace $\mathcal{A}$ of $\mathrm{Mat}_{n\times n}(\mathbb{C})$ is said to be a coherent algebra of order n if:

- $I, J \in \mathcal{A}$.
- $M^T \in \mathcal{A}$ for all $M \in \mathcal{A}$.
- $MN \in \mathcal{A}$ and $M \circ N \in \mathcal{A}$ for all $M, N \in \mathcal{A}$.

A coherent algebra $\mathcal{A}$ is said to be:

- Homogeneous if every matrix in $\mathcal{A}$ has a constant diagonal.
- Commutative if $\mathcal{A}$ is commutative with respect to ordinary matrix multiplication.
- Symmetric if every matrix in $\mathcal{A}$ is symmetric.

The set $\Gamma(\mathcal{A})$ of Schur-primitive matrices in a coherent algebra $\mathcal{A}$ is defined as,

$$\Gamma(\mathcal{A}) := \{M \in \mathcal{A} : M \circ M = M, M \circ N \in \mathrm{span}\{M\} \text{ for all } N \in \mathcal{A}\}.$$

Dually, the set $\Lambda(\mathcal{A})$ of primitive matrices in a coherent algebra $\mathcal{A}$ is defined as

$$\Lambda(\mathcal{A}) := \{M \in \mathcal{A} : M^2 = M, MN \in \mathrm{span}\{M\} \text{ for all } N \in \mathcal{A}\}.$$

Examples:

- The centralizer of a group of permutation matrices is a coherent algebra, i.e. $\mathcal{W}$ is a coherent algebra of order n if $\mathcal{W} := \{M \in \mathrm{Mat}_{n\times n}(\mathbb{C}) : MP = PM \text{ for all } P \in S\}$ for a group S of $n \times n$ permutation matrices. Additionally, the centralizer of the group of permutation matrices representing the automorphism group of a graph G is homogeneous if and only if G is vertex-transitive.
- The span of the set of matrices relating pairs of elements lying in the same orbit of a diagonal action of a finite group on a finite set is a coherent algebra, i.e. $\mathcal{W} := \mathrm{span}\{A(u,v) : u, v \in V\}$ where $A(u,v) \in \mathrm{Mat}_{V\times V}(\mathbb{C})$ is defined as $(A(u,v))_{x,y} := \begin{cases} 1 \text{ if } (x,y) = (u^g, v^g) \text{ for some } g \in G \\ 0 \text{ otherwise} \end{cases}$ for all $u, v \in V$ of a finite set V acted on by a finite group G.
- The span of a regular representation of a finite group as a group of permutation matrices over $\mathbb{C}$ is a coherent algebra.

Properties

- The intersection of a set of coherent algebras of order n is a coherent algebra.
- The tensor product of coherent algebras is a coherent algebra, i.e. $\mathcal{A}\otimes\mathcal{B}:=\{M\otimes N : M\in\mathcal{A} \text{ and } N\in\mathcal{B}\}$ if $\mathcal{A}\in \mathrm{Mat}_{m\times m}(\mathbb{C})$ and $\mathcal{B}\in \mathrm{Mat}_{n\times n}(\mathbb{C})$ are coherent algebras.
- The symmetrization $\widehat{\mathcal{A}}:=\mathrm{span}\{M+M^T : M\in\mathcal{A}\}$ of a commutative coherent algebra $\mathcal{A}$ is a coherent algebra.
- If $\mathcal{A}$ is a coherent algebra, then $M^T\in\Gamma(\mathcal{A})$ for all $M\in\mathcal{A}$, $\mathcal{A}=\mathrm{span}(\Gamma(\mathcal{A}))$, and $I\in\tilde{\Lambda}(\mathcal{A})$ if $\mathcal{A}$ is homogeneous.
- Dually, if $\mathcal{A}$ is a commutative coherent algebra (of order n), then $E^T, E^*\in\Lambda(\mathcal{A})$ for all $E\in\mathcal{A}$, $\frac{1}{n}J\in\Lambda(\mathcal{A})$, and $\mathcal{A}=\mathrm{span}(\Lambda(\mathcal{A}))$ as well.
- Every symmetric coherent algebra is commutative, and every commutative coherent algebra is homogeneous.
- A coherent algebra is commutative if and only if it is the Bose–Mesner algebra of a (commutative) association scheme.
- A coherent algebra forms a principal ideal ring under Schur product; moreover, a commutative coherent algebra forms a principal ideal ring under ordinary matrix multiplication as well.

BENDER–KNUTH INVOLUTION

In algebraic combinatorics, a Bender–Knuth involution is an involution on the set of semistandard tableaux, introduced by Bender & Knuth in their study of plane partitions. The Bender–Knuth involutions σ_k are defined for integers k, and act on the set of semistandard skew Young tableaux of some fixed shape μ/ν, where μ and ν are partitions. It acts by changing some of the elements k of the tableau to k + 1, and some of the entries k + 1 to k, in such a way that the numbers of elements with values k or k + 1 are exchanged. Call an entry of the tableau free if it is k or k + 1 and there is no other element with value k or k + 1 in the same column. For any i, the free entries of row i are all in consecutive columns, and consist of a_i copies of k followed by b_i copies of k + 1, for some a_i and b_i. The Bender–Knuth involution σ_k replaces them by b_i copies of k followed by a_i copies of k + 1.

Applications

Bender–Knuth involutions can be used to show that the number of semistandard skew tableaux of given shape and weight is unchanged under permutations of the weight. In turn this implies that the Schur function of a partition is a symmetric function.

H-VECTOR

In algebraic combinatorics, the h-vector of a simplicial polytope is a fundamental invariant of the polytope which encodes the number of faces of different dimensions and allows one to express the Dehn–Sommerville equations in a particularly simple form. A characterization of the set of h-vectors of simplicial polytopes was conjectured by Peter McMullen and proved by Lou Billera and Carl W. Lee and Richard Stanley (g-theorem). The definition of h-vector applies to arbitrary abstract simplicial complexes. The g-conjecture stated that for simplicial spheres, all possible h-vectors occur already among the h-vectors of the boundaries of convex simplicial polytopes. It was proven in December 2018 by Karim Adiprasito.

Stanley introduced a generalization of the h-vector, the toric h-vector, which is defined for an arbitrary ranked poset, and proved that for the class of Eulerian posets, the Dehn–Sommerville equations continue to hold. A different, more combinatorial, generalization of the h-vector that has been extensively studied is the flag h-vector of a ranked poset. For Eulerian posets, it can be more concisely expressed by means of a noncommutative polynomial in two variables called the cd-index.

Let Δ be an abstract simplicial complex of dimension $d - 1$ with f_i i-dimensional faces and $f_{-1} = 1$. These numbers are arranged into the f-vector of Δ,

$$f(\Delta) = (f_{-1}, f_0, \ldots, f_{d-1}).$$

An important special case occurs when Δ is the boundary of a *d*-dimensional convex polytope.

For $k = 0, 1, \ldots, d$, let,

$$h_k = \sum_{i=0}^{k} (-1)^{k-i} \binom{d-i}{k-i} f_{i-1}.$$

The tuple,

$$h(\Delta) = (h_0, h_1, \ldots, h_d)$$

is called the **h**-vector of Δ. The *f*-vector and the *h*-vector uniquely determine each other through the linear relation,

$$\sum_{i=0}^{d} f_{i-1}(t-1)^{d-i} = \sum_{k=0}^{d} h_k t^{d-k}.$$

Let $R = k[\Delta]$ be the Stanley–Reisner ring of Δ. Then its Hilbert–Poincaré series can be expressed as

$$P_R(t) = \sum_{i=0}^{d} \frac{f_{i-1} t^i}{(1-t)^i} = \frac{h_0 + h_1 t + \cdots + h_d t^d}{(1-t)^d}.$$

This motivates the definition of the h-vector of a finitely generated positively graded algebra of Krull dimension d as the numerator of its Hilbert–Poincaré series written with the denominator $(1 - t)^d$. The h-vector is closely related to the h^*-vector for a convex lattice polytope.

Toric h-vector

To an arbitrary graded poset P, Stanley associated a pair of polynomials f(P,x) and g(P,x). Their definition is recursive in terms of the polynomials associated to intervals [0,y] for all $y \in P, y \neq 1$, viewed as ranked posets of lower rank (0 and 1 denote the minimal and the maximal elements of P). The coefficients of f(P,x) form the toric h-vector of P. When P is an Eulerian poset of rank d + 1 such that P – 1 is simplicial, the toric h-vector coincides with the ordinary h-vector constructed using the numbers f_i of elements of P – 1 of given rank i + 1. In this case the toric h-vector of P satisfies the Dehn–Sommerville equations:

$$h_k = h_{d-k}.$$

The reason for the adjective "toric" is a connection of the toric h-vector with the intersection cohomology of a certain projective toric variety X whenever P is the boundary complex of rational convex polytope. Namely, the components are the dimensions of the even intersection cohomology groups of X:

$$h_k = \dim_{\mathbb{Q}} IH^{2k}(X, \mathbb{Q})$$

(the odd intersection cohomology groups of X are all zero). The Dehn–Sommerville equations are a manifestation of the Poincaré duality in the intersection cohomology of X. Kalle Karu proved that the toric h-vector of a polytope is unimodal, regardless of whether the polytope is rational or not.

Flag h-vector and cd-index

A different generalization of the notions of f-vector and h-vector of a convex polytope has been extensively studied. Let P be a finite graded poset of rank n, so that each maximal chain in P has length n. For any S, a subset of $\{0,\ldots,n\}$, let $\alpha_P(S)$ denote the number of chains in P whose ranks constitute the set S. More formally, let:

$$rk : P \to \{0,1,\ldots,n\}$$

be the rank function of P and let P_S be the S – rank selected subposet, which consists of the elements from P whose rank is in S:

$$P_S = \{x \in P : rk(x) \in S\}.$$

Then $\alpha_P(S)$ is the number of the maximal chains in P_S and the function:

$$S \mapsto \alpha_P(S)$$

is called the flag f-vector of P. The function,

$$S \mapsto \beta_P(S), \quad \beta_P(S) = \sum_{T \subseteq S} (-1)^{|S|-|T|} \alpha_P(S)$$

is called the flag *h*-vector of P. By the inclusion–exclusion principle,

$$\alpha_P(S) = \sum_{T \subseteq S} \beta_P(T).$$

The flag f- and h-vectors of P refine the ordinary f- and h-vectors of its order complex $\Delta(P)$:

$$f_{i-1}(\Delta(P)) = \sum_{|S|=i} \alpha_P(S), \quad h_i(\Delta(P)) = \sum_{|S|=i} \beta_P(S).$$

The flag h-vector of P can be displayed via a polynomial in noncommutative variables a and b. For any subset *S* of {1,...,n}, define the corresponding monomial in a and b,

$$u_S = u_1 \cdots u_n, \quad u_i = a \text{ for } i \notin S, u_i = b \text{ for } i \in S.$$

Then the noncommutative generating function for the flag h-vector of P is defined by,

$$\Psi_P(a,b) = \sum_S \beta_P(S) u_S.$$

From the relation between $\alpha_P(S)$ and $\beta_P(S)$, the noncommutative generating function for the flag f-vector of P is,

$$\Psi_P(a, a+b) = \sum_S \alpha_P(S) u_S.$$

Margaret Bayer and Lou Billera determined the most general linear relations that hold between the components of the flag h-vector of an Eulerian poset P.

Fine noted an elegant way to state these relations: there exists a noncommutative polynomial $\Phi_P(c,d)$, called the cd-index of P, such that,

$$\Psi_P(a,b) = \Phi_P(a+b, ab+ba).$$

Stanley proved that all coefficients of the cd-index of the boundary complex of a convex polytope are non-negative. He conjectured that this positivity phenomenon persists for a more general class of Eulerian posets that Stanley calls Gorenstein* complexes and which includes simplicial spheres and complete fans. This conjecture was proved by Kalle Karu.

SIMPLICIAL SPHERE

In geometry and combinatorics, a simplicial (or combinatorial) *d*-sphere is a simplicial complex homeomorphic to the *d*-dimensional sphere. Some simplicial spheres arise as the boundaries of convex polytopes, however, in higher dimensions most simplicial spheres cannot be obtained in this way.

One important open problem in the field was the g-conjecture, formulated by Peter McMullen, which asks about possible numbers of faces of different dimensions of a simplicial sphere.

Examples:

- For any n ≥ 3, the simple n-cycle C_n is a simplicial circle, i.e. a simplicial sphere of dimension 1. This construction produces all simplicial circles.
- The boundary of a convex polyhedron in R^3 with triangular faces, such as an octahedron or icosahedron, is a simplicial 2-sphere.
- More generally, the boundary of any (d+1)-dimensional compact (or bounded) simplicial convex polytope in the Euclidean space is a simplicial d-sphere.

Properties

It follows from Euler's formula that any simplicial 2-sphere with n vertices has 3n – 6 edges and 2n – 4 faces. The case of n = 4 is realized by the tetrahedron. By repeatedly performing the barycentric subdivision, it is easy to construct a simplicial sphere for any n ≥ 4. Moreover, Ernst Steinitz gave a characterization of 1-skeleta (or edge graphs) of convex polytopes in R^3 implying that any simplicial 2-sphere is a boundary of a convex polytope. Branko Grünbaum constructed an example of a non-polytopal simplicial sphere (that is, a simplicial sphere that is not the boundary of a polytope). Gil Kalai proved that, in fact, "most" simplicial spheres are non-polytopal. The smallest example is of dimension d = 4 and has $f_0 = 8$ vertices.

The upper bound theorem gives upper bounds for the numbers *fi* of *i*-faces of any simplicial *d*-sphere with $f_0 = n$ vertices. This conjecture was proved for polytopal spheres by Peter McMullen in 1970 and by Richard Stanley for general simplicial spheres.

The g-conjecture, formulated by McMullen in 1970, asks for a complete characterization of f-vectors of simplicial d-spheres. In other words, what are the possible sequences of numbers of faces of each dimension for a simplicial d-sphere? In the case of polytopal spheres, the answer is given by the g-theorem, proved in 1979 by Billera and Lee (existence) and Stanley (necessity). It has been conjectured that the same conditions are necessary for general simplicial spheres.

RING OF SYMMETRIC FUNCTIONS

In algebra and in particular in algebraic combinatorics, the ring of symmetric functions is a specific limit of the rings of symmetric polynomials in n indeterminates, as n goes to infinity. This ring serves as universal structure in which relations between symmetric polynomials can be expressed in a way independent of the number n of indeterminates (but its elements are neither polynomials nor functions). Among other things, this ring plays an important role in the representation theory of the symmetric group. The ring of symmetric functions can be given a coproduct and a bilinear form making it into a positive selfadjoint graded Hopf algebra that is both commutative and cocommutative.

Symmetric Polynomials

The study of symmetric functions is based on that of symmetric polynomials. In a polynomial ring in some finite set of indeterminates, a polynomial is called symmetric if it stays the same whenever the indeterminates are permuted in any way. More formally, there is an action by ring automorphisms of the symmetric group S_n on the polynomial ring in n indeterminates, where a permutation acts on a polynomial by simultaneously substituting each of the indeterminates for another according to the permutation used. The invariants for this action form the subring of symmetric polynomials. If the indeterminates are $X_1,...,X_n$, then examples of such symmetric polynomials are:

$$X_1 + X_2 + \cdots + X_n,$$

$$X_1^3 + X_2^3 + \cdots + X_n^3,$$

and

$$X_1 X_2 \cdots X_n.$$

A somewhat more complicated example is $X_1^3X_2X_3 + X_1X_2^3X_3 + X_1X_2X_3^3 + X_1^3X_2X_4 + X_1X_2^3X_4 + X_1X_2X_4^3 + ...$ where the summation goes on to include all products of the third power of some variable and two other variables. There are many specific kinds of symmetric polynomials, such as elementary symmetric polynomials, power sum symmetric polynomials, monomial symmetric polynomials, complete homogeneous symmetric polynomials, and Schur polynomials.

Ring of Symmetric Functions

Most relations between symmetric polynomials do not depend on the number n of indeterminates, other than that some polynomials in the relation might require n to be large enough in order to be defined. For instance the Newton's identity for the third power sum polynomial p_3 leads to,

$$p_3(X_1,...,X_n) = e_1(X_1,...,X_n)^3 - 3e_2(X_1,...,X_n)e_1(X_1,...,X_n) + 3e_3(X_1,...,X_n),$$

where the e_i denote elementary symmetric polynomials; this formula is valid for all natural numbers n, and the only notable dependency on it is that $e_k(X_1,...,X_n) = 0$ whenever $n < k$. One would like to write this as an identity,

$$p_3 = e_1^3 - 3e_2e_1 + 3e_3$$

that does not depend on n at all, and this can be done in the ring of symmetric functions. In that ring there are elements e_k for all integers $k \geq 1$, and any element of the ring can be given by a polynomial expression in the elements e_k.

A ring of symmetric functions can be defined over any commutative ring R, and will be denoted Λ_R; the basic case is for R = Z. The ring Λ_R is in fact a graded R-algebra.

Ring of Formal Power Series

The easiest (though somewhat heavy) construction starts with the ring of formal power series $R[[X_1, X_2, \ldots]]$ over *R* in infinitely (countably) many indeterminates; the elements of this power series ring are formal infinite sums of terms, each of which consists of a coefficient from *R* multiplied by a monomial, where each monomial is a product of finitely many finite powers of indeterminates. One defines Λ*R* as its subring consisting of those power series *S* that satisfy:

- *S* is invariant under any permutation of the indeterminates.
- The degrees of the monomials occurring in *S* are bounded.

Because of the second condition, power series are used here only to allow infinitely many terms of a fixed degree, rather than to sum terms of all possible degrees. Allowing this is necessary because an element that contains for instance a term X_1 should also contain a term X_i for every $i > 1$ in order to be symmetric. Unlike the whole power series ring, the subring Λ_R is graded by the total degree of monomials: due to condition 2, every element of Λ_R is a finite sum of homogeneous elements of Λ_R (which are themselves infinite sums of terms of equal degree). For every $k \geq 0$, the element $e_k \in \Lambda_R$ is defined as the formal sum of all products of k distinct indeterminates, which is clearly homogeneous of degree k.

Algebraic Limit

Another construction of Λ_R takes somewhat longer to describe, but better indicates the relationship with the rings $R[X_1,\ldots,X_n]^S{}_n$ of symmetric polynomials in n indeterminates. For every n there is a surjective ring homomorphism ρ_n from the analogous ring $R[X_1,\ldots,X_{n+1}]^S{}_{n+1}$ with one more indeterminate onto $R[X_1,\ldots,X_n]^S{}_n$, defined by setting the last indeterminate X_{n+1} to 0. Although ρ_n has a non-trivial kernel, the nonzero elements of that kernel have degree at least $n+1$ (they are multiples of $X_1X_2\ldots X_{n+1}$). This means that the restriction of ρ_n to elements of degree at most n is a bijective linear map, and $\rho_n(e_k(X_1,\ldots,X_{n+1})) = e_k(X_1,\ldots,X_n)$ for all $k \leq n$. The inverse of this restriction can be extended uniquely to a ring homomorphism φ_n from $R[X_1,\ldots,X_n]^S{}_n$ to $R[X_1,\ldots,X_{n+1}]^S{}_{n+1}$, as follows for instance from the fundamental theorem of symmetric polynomials. Since the images $\varphi_n(e_k(X_1,\ldots,X_n)) = e_k(X_1,\ldots,X_{n+1})$ for $k = 1,\ldots,n$ are still algebraically independent over R, the homomorphism φ_n is injective and can be viewed as a (somewhat unusual) inclusion of rings; applying φ_n to a polynomial amounts to adding all monomials containing the new indeterminate obtained by symmetry from monomials already present. The ring Λ_R is then the "union" (direct limit) of all these rings subject to these inclusions. Since all φ_n are compatible with the grading by total degree of the rings involved, Λ_R obtains the structure of a graded ring.

That construction only uses the surjective morphisms ρn without mentioning the injective morphisms φn: it constructs the homogeneous components of ΛR separately, and equips their direct sum with a ring structure using the ρn. It is also observed that the result can be described as an inverse limit in the category of graded rings. That description however somewhat obscures an important property typical for a direct limit of injective morphisms, namely that every individual element (symmetric function) is already faithfully represented in some object used in the limit construction, here a ring $R[X_1,\ldots,X_d]^S{}_d$. It suffices to take for d the degree of the symmetric function, since the part in degree d of that ring is mapped isomorphically to rings with more indeterminates by φ_n for all $n \geq d$. This implies that for studying relations between individual elements, there is no fundamental difference between symmetric polynomials and symmetric functions.

Defining Individual Symmetric Functions

The name "symmetric function" for elements of ΛR is a misnomer: in neither construction the elements are functions, and in fact, unlike symmetric polynomials, no function of independent variables can be associated to such elements (for instance e_1 would be the sum of all infinitely many variables, which is not defined unless restrictions are imposed on the variables). However the name is traditional and well established; it can be found both in, which says:

> "The elements of Λ (unlike those of Λn) are no longer polynomials: they are formal infinite sums of monomials. We have therefore reverted to the older terminology of symmetric functions."

(here Λn denotes the ring of symmetric polynomials in n indeterminates).

To define a symmetric function one must either indicate directly a power series as in the first construction, or give a symmetric polynomial in n indeterminates for every natural number n in a way compatible with the second construction. An expression in an unspecified number of indeterminates may do both, for instance:

$$e_2 = \sum_{i<j} X_i X_j$$

can be taken as the definition of an elementary symmetric function if the number of indeterminates is infinite, or as the definition of an elementary symmetric polynomial in any finite number of indeterminates. Symmetric polynomials for the same symmetric function should be compatible with the morphisms ρn (decreasing the number of indeterminates is obtained by setting some of them to zero, so that the coefficients of any monomial in the remaining indeterminates is unchanged), and their degree should remain bounded. (An example of a family of symmetric polynomials that fails both conditions is $\Pi_{i=1}^{n} X_i$; the family $\Pi_{i=1}^{n}(X_i + 1)$ fails only the second condition). Any symmetric polynomial in n indeterminates can be used to construct a compatible family of symmetric polynomials, using the morphisms ρ_i for $i < n$ to decrease the number of indeterminates, and φ_i for $i \geq n$ to increase the number of indeterminates (which amounts to adding all monomials in new indeterminates obtained by symmetry from monomials already present).

The following are fundamental examples of symmetric functions:

- The monomial symmetric functions m_α. Suppose $\alpha = (\alpha_1, \alpha_2, \ldots)$ is a sequence of non-negative integers, only finitely many of which are non-zero. Then we can consider the monomial defined by α: $X^\alpha = X_1^{\alpha_1} X_2^{\alpha_2} X_3^{\alpha_3} \cdots$. Then m_α is the symmetric function determined by X^α, i.e. the sum of all monomials obtained from X^α by symmetry. For a formal definition, define $\beta \sim \alpha$ to mean that the sequence β is a permutation of the sequence α and set:

  $$m_\alpha = \sum_{\beta \sim \alpha} X^\beta.$$

  This symmetric function corresponds to the monomial symmetric polynomial $m_\alpha(X_1, \ldots, X_n)$ for any n large enough to have the monomial X^α. The distinct monomial symmetric functions are parametrized by the integer partitions (each m_α has a unique representative monomial X^λ with the parts λ_i in weakly decreasing order). Since any symmetric function containing any of the monomials of some m_α must contain all of them with the same coefficient, each symmetric function can be written as an R-linear combination of monomial symmetric functions, and the distinct monomial symmetric functions therefore form a basis of Λ_R as R-module.

- The elementary symmetric functions e_k, for any natural number k; one has $e_k = m_\alpha$ where $X^\alpha = \Pi_{i=1}^{k} X_i$. As a power series, this is the sum of all distinct products of k distinct indeterminates. This symmetric function corresponds to the elementary symmetric polynomial $e_k(X_1, \ldots, X_n)$ for any $n \geq k$.

- The power sum symmetric functions p_k, for any positive integer k; one has $p_k = m_{(k)}$, the monomial symmetric function for the monomial X_1^k. This symmetric function corresponds to the power sum symmetric polynomial $p_k(X_1, \ldots, X_n) = X_1^k + \ldots + X_n^k$ for any $n \geq 1$.

- The complete homogeneous symmetric functions h_k, for any natural number k; h_k is the sum of all monomial symmetric functions m_α where α is a partition of k. As a power series, this is the sum of all monomials of degree k, which is what motivates its name. This symmetric function corresponds to the complete homogeneous symmetric polynomial $h_k(X_1, \ldots, X_n)$ for any $n \geq k$.

- The Schur functions s_λ for any partition λ, which corresponds to the Schur polynomial $s_\lambda(X_1, \ldots, X_n)$ for any n large enough to have the monomial X^λ.

There is no power sum symmetric function p_0: although it is possible (and in some contexts natural) to define $p_0(X_1,\dots,X_n) = \Sigma_{i=1}^{n} X_i^0 = n$ as a symmetric polynomial in n variables, these values are not compatible with the morphisms ρn. The "discriminant" $(\prod_{i<j}(X_i - X_j))^2$ is another example of an expression giving a symmetric polynomial for all n, but not defining any symmetric function. The expressions defining Schur polynomials as a quotient of alternating polynomials are somewhat similar to that for the discriminant, but the polynomials $s_\lambda(X_1,\dots,X_n)$ turn out to be compatible for varying n, and therefore do define a symmetric function.

A Principle Relating Symmetric Polynomials and Symmetric Functions

For any symmetric function P, the corresponding symmetric polynomials in n indeterminates for any natural number n may be designated by $P(X_1,\dots,X_n)$. The second definition of the ring of symmetric functions implies the following fundamental principle:

> "If P and Q are symmetric functions of degree d, then one has the identity $P = Q$ of symmetric functions if and only one has the identity $P(X_1,\dots,X_d) = Q(X_1,\dots,X_d)$ of symmetric polynomials in d indeterminates. In this case one has in fact $P(X_1,\dots,X_n) = Q(X_1,\dots,X_n)$ for any number n of indeterminates."

This is because one can always reduce the number of variables by substituting zero for some variables, and one can increase the number of variables by applying the homomorphisms φ_n; the definition of those homomorphisms assures that $\varphi_n(P(X_1,\dots,X_n)) = P(X_1,\dots,X_{n+1})$ (and similarly for Q) whenever $n \geq d$.

Properties of the Ring of Symmetric Functions

The ring of symmetric functions is a convenient tool for writing identities between symmetric polynomials that are independent of the number of indeterminates: in ΛR there is no such number, yet by the above principle any identity in ΛR automatically gives identities the rings of symmetric polynomials over R in any number of indeterminates. Some fundamental identities are:

$$\sum_{i=0}^{k}(-1)^i e_i h_{k-i} = 0 = \sum_{i=0}^{k}(-1)^i h_i e_{k-i} \quad \text{for all } k > 0,$$

which shows a symmetry between elementary and complete homogeneous symmetric functions; these relations are explained under complete homogeneous symmetric polynomial,

$$ke_k = \sum_{i=1}^{k}(-1)^{i-1} p_i e_{k-i} \quad \text{for all } k \geq 0,$$

the Newton identities, which also have a variant for complete homogeneous symmetric functions:

$$kh_k = \sum_{i=1}^{k} p_i h_{k-i} \quad \text{for all } k \geq 0.$$

Structural Properties of Λ_R

- Important properties of ΛR include the following.
- The set of monomial symmetric functions parametrized by partitions form a basis of Λ_R as graded R-module, those parametrized by partitions of d being homogeneous of degree d; the same is true for the set of Schur functions (also parametrized by partitions).
- Λ_R is isomorphic as a graded R-algebra to a polynomial ring $R[Y_1,Y_2,...]$ in infinitely many variables, where Y_i is given degree i for all i > 0, one isomorphism being the one that sends Y_i to $e_i \in \Lambda_R$ for every i.
- There is an involutory automorphism ω of Λ_R that interchanges the elementary symmetric functions e_i and the complete homogeneous symmetric function h_i for all i. It also sends each power sum symmetric function p_i to $(-1)^{i-1} p_i$, and it permutes the Schur functions among each other, interchanging s_λ and $s_\lambda{}^t$ where λ^t is the transpose partition of λ.

Property 2 is the essence of the fundamental theorem of symmetric polynomials. It immediately implies some other properties:

- The subring of ΛR generated by its elements of degree at most n is isomorphic to the ring of symmetric polynomials over R in n variables.
- The Hilbert–Poincaré series of ΛR is $\prod_{i=1}^{\infty} \frac{1}{1-t^i}$, the generating function of the integer partitions (this also follows from property 1).
- For every n > 0, the R-module formed by the homogeneous part of Λ_R of degree n, modulo its intersection with the subring generated by its elements of degree strictly less than n, is free of rank 1, and the image of e_n is a generator of this R-module.
- For every family of symmetric functions $(f_i)_{i>0}$ in which f_i is homogeneous of degree i and gives a generator of the free R-module of the previous point (for all i), there is an alternative isomorphism of graded R-algebras from $R[Y_1,Y_2,...]$ as above to Λ_R that sends Y_i to f_i; in other words, the family $(f_i)_{i>0}$ forms a set of free polynomial generators of Λ_R.

This final point applies in particular to the family $(h_i)_{i>0}$ of complete homogeneous symmetric functions. If R contains the field $\mathbb{Q}$ of rational numbers, it applies also to the family $(p_i)_{i>0}$ of power sum symmetric functions. This explains why the first n elements of each of these families define sets of symmetric polynomials in n variables that are free polynomial generators of that ring of symmetric polynomials.

The fact that the complete homogeneous symmetric functions form a set of free polynomial generators of ΛR already shows the existence of an automorphism ω sending the elementary symmetric functions to the complete homogeneous ones. The fact that ω is an involution of ΛR follows from the symmetry between elementary and complete homogeneous symmetric functions expressed by the first set of relations. The ring of symmetric functions Λ_Z is the Exp ring of the integers Z. It is also a lambda-ring in a natural fashion; in fact it is the universal lambda-ring in one generator.

Generating Functions

The first definition of Λ_R as a subring of $R[[X_1,X_2,...]]$ allows the generating functions of several sequences of symmetric functions to be elegantly expressed. Contrary to the relations, which are internal to Λ_R, these expressions involve operations taking place in $R[[X_1,X_2,...;t]]$ but outside its subring $\Lambda_R[[t]]$, so they are meaningful only if symmetric functions are viewed as formal power series in indeterminates X_i. We shall write "(X)" after the symmetric functions to stress this interpretation.

The generating function for the elementary symmetric functions is:

$$E(t)=\sum_{k\geq 0} e_k(X)t^k=\prod_{i=1}^{\infty}(1+X_i t).$$

Similarly one has for complete homogeneous symmetric functions:

$$H(t)=\sum_{k\geq 0} h_k(X)t^k=\prod_{i=1}^{\infty}\left(\sum_{k\geq 0}(X_i t)^k\right)=\prod_{i=1}^{\infty}\frac{1}{1-X_i t}.$$

The obvious fact that $E(-t)H(t)=1=E(t)H(-t)$ explains the symmetry between elementary and complete homogeneous symmetric functions. The generating function for the power sum symmetric functions can be expressed as:

$$P(t)=\sum_{k>0} p_k(X)t^k=\sum_{k>0}\sum_{i=1}^{\infty}(X_i t)^k=\sum_{i=1}^{\infty}\frac{X_i t}{1-X_i t}=\frac{tE'(-t)}{E(-t)}=\frac{tH'(t)}{H(t)}$$

defines P(t) as $\Sigma_{k>0}\, p_k(X)t^{k-1}$, and its expressions therefore lack a factor t with respect to those given here). The two final expressions, involving the formal derivatives of the generating functions E(t) and H(t), imply Newton's identities and their variants for the complete homogeneous symmetric functions. These expressions are sometimes written as,

$$P(t)=-t\frac{d}{dt}\log(E(-t))=t\frac{d}{dt}\log(H(t)),$$

which amounts to the same, but requires that *R* contain the rational numbers, so that the logarithm of power series with constant term 1 is defined (by $\log(1-tS) = -\sum_{i>0} \frac{1}{i}(tS)^i$).

STANLEY'S RECIPROCITY THEOREM

In combinatorial mathematics, Stanley's reciprocity theorem, named after MIT mathematician Richard P. Stanley, states that a certain functional equation is satisfied by the generating function of any rational cone and the generating function of the cone's interior.

A rational cone is the set of all d-tuples,

$$(a_1,\ldots, a_d)$$

of nonnegative integers satisfying a system of inequalities,

$$M\begin{bmatrix} a_1 \\ \vdots \\ a_d \end{bmatrix} \geq \begin{bmatrix} 0 \\ \vdots \\ 0 \end{bmatrix}$$

where M is a matrix of integers. A d-tuple satisfying the corresponding strict inequalities, i.e., with ">" rather than "≥", is in the interior of the cone.

The generating function of such a cone is:

$$F(x_1,\ldots,x_d) = \sum_{(a_1,\ldots,a_d)\in \text{cone}} x_1^{a_1} \cdots x_d^{a_d}.$$

The generating function $F_{\text{int}}(x_1,\ldots, x_d)$ of the interior of the cone is defined in the same way, but one sums over d-tuples in the interior rather than in the whole cone. These are rational functions.

Formulation

Stanley's reciprocity theorem states that for a rational cone as above, we have

$$F(1/x_1,\ldots,1/x_d) = (-1)^d F_{\text{int}}(x_1,\ldots,x_d).$$

Matthias Beck and Mike Develin have shown how to prove this by using the calculus of residues. Develin has said that this amounts to proving the result "without doing any work". Stanley's reciprocity theorem generalizes Ehrhart-Macdonald reciprocity for Ehrhart polynomials of rational convex polytopes.

QUASISYMMETRIC FUNCTION

In algebra and in particular in algebraic combinatorics, a quasisymmetric function is any element in the ring of quasisymmetric functions which is in turn a subring of the formal power series ring with a countable number of variables. This ring generalizes the ring of symmetric functions. This ring can be realized as a specific limit of the rings of quasisymmetric polynomials in n variables, as n goes to infinity. This ring serves as universal structure in which relations between quasisymmetric polynomials can be expressed in a way independent of the number n of variables (but its elements are neither polynomials nor functions).

The ring of quasisymmetric functions, denoted QSym, can be defined over any commutative ring *R* such as the integers. Quasisymmetric functions are power series of bounded degree in variables $x_1, x_2, x_3, \ldots$ with coefficients in *R*, which are shift invariant in the sense that the coefficient of the monomial $x_1^{\alpha_1} x_2^{\alpha_2} \cdots x_k^{\alpha_k}$ is equal to the coefficient of the monomial $x_{i_1}^{\alpha_1} x_{i_2}^{\alpha_2} \cdots x_{i_k}^{\alpha_k}$ for any strictly increasing sequence of positive integers $i_1 < i_2 < \cdots < i_k$ indexing the variables and any positive integer sequence $(\alpha_1, \alpha_2, \ldots, \alpha_k)$ of exponents. Much of the study of quasisymmetric functions is based on that of symmetric functions.

A quasisymmetric function in finitely many variables is a *quasisymmetric polynomial*. Both symmetric and quasisymmetric polynomials may be characterized in terms of actions of the symmetric group S_n on a polynomial ring in n variables $x_1, \ldots, x_n$. One such action of S_n permutes variables, changing a polynomial $p(x_1, \ldots, x_n)$ by iteratively swapping pairs (x_i, x_{i+1}) of variables having consecutive indices. Those polynomials unchanged by all such swaps form the subring of symmetric polynomials. A second action of S_n conditionally permutes variables, changing a polynomial $p(x_1, \ldots, x_n)$ by swapping pairs (x_i, x_{i+1}) of variables except in monomials containing both variables. Those polynomials unchanged by all such conditional swaps form the subring of quasisymmetric polynomials. One quasisymmetric function in four variables is the polynomial:

$$x_1^2 x_2 x_3 + x_1^2 x_2 x_4 + x_1^2 x_3 x_4 + x_2^2 x_3 x_4.$$

The simplest symmetric function containing all of these monomials is:

$$x_1^2 x_2 x_3 + x_1^2 x_2 x_4 + x_1^2 x_3 x_4 + x_2^2 x_3 x_4 + x_1 x_2^2 x_3 + x_1 x_2^2 x_4 + x_1 x_3^2 x_4 + x_2 x_3^2 x_4 + x_1 x_2 x_3^2 + x_1 x_2 x_4^2 + x_1 x_3 x_4^2 + x_2 x_3 x_4^2.$$

Important Bases

QSym is a graded R-algebra, decomposing as,

$$\mathrm{QSym} = \bigoplus_{n \geq 0} \mathrm{QSym}_n,$$

where QSym is the R – span of all quasisymmetric functions that are homogeneous of degree n. Two natural bases for $QSym_n$ are the monomial basis $\{M_\alpha\}$ and the fundamental basis $\{F_{\acute{a}}\}$ indexed by compositions $\alpha = (\alpha_1, \alpha_2, \ldots, \alpha_k)$ of n, denoted $\alpha \models n$. The monomial basis consists of $M_0 = 1$ and all formal power series:

$$M_\alpha = \sum_{i_1 < i_2 < \cdots < i_k} x_{i_1}^{\alpha_1} x_{i_2}^{\alpha_2} \cdots x_{i_k}^{\alpha_k}.$$

The fundamental basis consists $F_0 = 1$ and all formal power series,

$$F_\alpha = \sum_{\alpha \geq \beta} M_\beta,$$

where $\alpha \succeq \beta$ means we can obtain α by adding together adjacent parts of $\hat{a}$, for example, (3,2,4,2) $\succeq$ (3, 1, 1, 1, 2, 1, 2). Thus, when the ring R is the ring of rational numbers, one has,

$$QSym_n = span_Q\{M_\alpha \mid \alpha \models n\} = span_Q\{F_\alpha \mid \alpha \models n\}.$$

Then one can define the algebra of symmetric functions $\Lambda = \Lambda_0 \oplus \Lambda_1 \oplus \cdots$ as the subalgebra of QSym spanned by the monomial symmetric functions $m_0 = 1$ and all formal power series $m_\lambda = \sum M_\alpha$, where the sum is over all compositions α which rearrange to the partition $\ddot{e}$. Moreover, we have $\Lambda_n = \Lambda \cap QSym_n$. For example, $F_{(1,2)} = M_{(1,2)} + M_{(1,1,1)}$ and $m_{(2,1)} = M_{(2,1)} + M_{(1,2)}$. Other important bases for quasisymmetric functions include the basis of quasisymmetric Schur functions, and bases related to enumeration in matroids.

Applications

Quasisymmetric functions have been applied in enumerative combinatorics, symmetric function theory, representation theory, and number theory. Applications of quasisymmetric functions include enumeration of P-partitions, permutations, tableaux, chains of posets, reduced decompositions in finite Coxeter groups (via Stanley symmetric functions), and parking functions. In symmetric function theory and representation theory, applications include the study of Schubert polynomials, Macdonald polynomials, Hecke algebras, and Kazhdan–Lusztig polynomials. Often quasisymmetric functions provide a powerful bridge between combinatorial structures and symmetric functions.

Related Algebras

As a graded Hopf algebra, the dual of the ring of quasisymmetric functions is the ring of noncommutative symmetric functions. Every symmetric function is also a quasisymmetric function, and hence the ring of symmetric functions is a subalgebra of the ring of quasisymmetric functions. The ring of quasisymmetric functions is the terminal object in category of graded Hopf algebras with a single character. Hence any such Hopf algebra has a morphism to the ring of quasisymmetric functions. One example of this is the peak algebra.

Other Related Algebras

The Malvenuto–Reutenauer algebra is a Hopf algebra based on permutations that relates the rings of symmetric functions, quasisymmetric functions, and noncommutative symmetric functions, (denoted Sym, QSym, and NSym respectively), as depicted the following commutative diagram. The duality between QSym and NSym mentioned above is reflected in the main diagonal of this diagram.

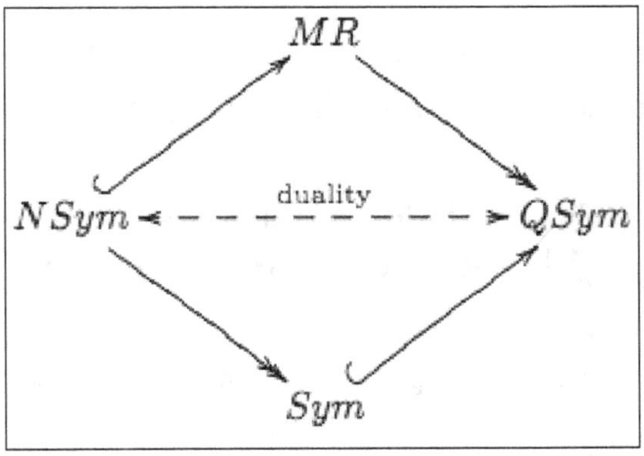

Many related Hopf algebras were constructed from Hopf monoids in the category of species by Aguiar and Majahan. One can also construct the ring of quasisymmetric functions in noncommuting variables.

EULERIAN POSET

In combinatorial mathematics, an Eulerian poset is a graded poset in which every nontrivial interval has the same number of elements of even rank as of odd rank. An Eulerian poset which is a lattice is an Eulerian lattice. These objects are named after Leonhard Euler. Eulerian lattices generalize face lattices of convex polytopes and much recent research has been devoted to extending known results from polyhedral combinatorics, such as various restrictions on *f*-vectors of convex simplicial polytopes, to this more general setting.

Examples:

- The face lattice of a convex polytope, consisting of its faces, together with the smallest element, the empty face, and the largest element, the polytope itself, is an Eulerian lattice. The odd–even condition follows from Euler's formula.
- Any simplicial generalized homology sphere is an Eulerian lattice.

- Let L be a regular cell complex such that |L| is a manifold with the same Euler characteristic as the sphere of the same dimension (this condition is vacuous if the dimension is odd). Then the poset of cells of L, ordered by the inclusion of their closures, is Eulerian.
- Let W be a Coxeter group with Bruhat order. Then (W,≤) is an Eulerian poset.

Properties

- The defining condition of an Eulerian poset P can be equivalently stated in terms of its Möbius function:

  $\mu_P(x,y) = (-1)^{|y|-|x|}$ for all $x \leq y$.
- The dual of an Eulerian poset, obtained by reversing the partial order, is Eulerian.
- Richard Stanley defined the toric h-vector of a ranked poset, which generalizes the h-vector of a simplicial polytope. He proved that the Dehn–Sommerville equations.

  $h_k = h_{d-k}$ hold for an arbitrary Eulerian poset of rank d + 1. However, for an Eulerian poset arising from a regular cell complex or a convex polytope, the toric h-vector neither determines, nor is neither determined by the numbers of the cells or faces of different dimension and the toric *h*-vector does not have a direct combinatorial interpretation.

BOSE–MESNER ALGEBRA

In mathematics, a Bose–Mesner algebra is a special set of matrices which arise from a combinatorial structure known as an association scheme, together with the usual set of rules for combining (forming the products of) those matrices, such that they form an associative algebra, or, more precisely, a unitary commutative algebra. Among these rules are:

- The result of a product is also within the set of matrices.
- There is an identity matrix in the set.
- Taking products is commutative.

Bose–Mesner algebras have applications in physics to spin models, and in statistics to the design of experiments. They are named for R. C. Bose and Dale Marsh Mesner.

Let X be a set of v elements. Consider a partition of the 2-element subsets of X into n non-empty subsets, $R_1,\ldots, R_n$ such that:

- Given an $x \in X$, the number of $y \in X$ such that $\{x,y\} \in R_i$ depends only on i (and not on x). This number will be denoted by v_i.

- Given $x,y \in X$ with $\{x,y\} \in R_k$, the number of $z \in X$ such that $\{x,z\} \in R_i$ and $\{z,y\} \in R_j$ depends only on i,j and k (and not on x and y). This number will be denoted by p^k_{ij}.

This structure is enhanced by adding all pairs of repeated elements of X and collecting them in a subset R_0. This enhancement permits the parameters i, j, and k to take on the value of zero, and lets some of x, y or z be equal.

A set with such an enhanced partition is called an association scheme. One may view an association scheme as a partition of the edges of a complete graph (with vertex set X) into n classes, often thought of as color classes. In this representation, there is a loop at each vertex and all the loops receive the same 0th color.

The association scheme can also be represented algebraically. Consider the matrices D_i defined by:

$$(D_i)_{x,y} = \begin{cases} 1, & \text{if } (x,y) \in R_i, \\ 0, & \text{otherwise.} \end{cases}$$

Let $\mathcal{A}$ be the vector space consisting of all matrices $\sum_{i=0}^{n} a_i D_i$, with a_i complex.

The definition of an association scheme is equivalent to saying that the D_i are $v \times v$ (0,1)-matrices which satisfy,

- D_i is symmetric,
- $\sum_{i=0}^{n} D_i = J$ (the all-ones matrix),
- $D_0 = I$,
- $D_i D_j = \sum_{k=0}^{n} p^k_{ij} D_k = D_j D_i, \qquad i,j = 0,\ldots,n.$

The (x,y)-th entry of the left side of 4. is the number of two colored paths of length two joining x and y (using "colors" i and j) in the graph. The rows and columns of D_i contain v_i 1s:

$$D_i J = J D_i = v_i J.$$

From 1., these matrices are symmetric. From 2., $D_0,\ldots,D_n$ are linearly independent, and the dimension of $\mathcal{A}$ is n + 1. From 4., $\mathcal{A}$ is closed under multiplication, and multiplication is always associative. This associative commutative algebra $\mathcal{A}$ is called the Bose–Mesner algebra of the association scheme. Since the matrices in $\mathcal{A}$ are symmetric and commute with each other, they can be simultaneously diagonalized. This means that there is a matrix S such that to each $A \in \mathcal{A}$ there is a diagonal matrix Λ_A with $S^{-1}AS = \Lambda_A$. This means that $\mathcal{A}$ is semi-simple and has a unique basis of primitive idempotents $J_0,\ldots,J_n$.

These are complex n × n matrices satisfying,

$$J_i^2 = J_i, i = 0, \ldots, n,$$

$$J_i J_k = 0, i \neq k,$$

$$\sum_{i=0}^{n} J_i = I.$$

The Bose–Mesner algebra has two distinguished bases: The basis consisting of the adjacency matrices D_i, and the basis consisting of the irreducible idempotent matrices E_k. By definition, there exist well-defined complex numbers such that,

$$D_i = \sum_{k=0}^{n} p_i(k) E_k,$$

and

$$|X| E_k = \sum_{i=0}^{n} q_k(i) D_i.$$

The p-numbers $p_i(k)$, and the q-numbers $q_k(i)$, play a prominent role in the theory. They satisfy well-defined orthogonality relations. The p-numbers are the eigenvalues of the adjacency matrix D_i.

Theorem: The eigenvalues of $p_i(k)$ and $q_k(i)$, satisfy the orthogonality conditions:

$$\sum_{k=0}^{n} \mu_i p_i(k) p_\ell(k) = v v_i \delta_{i\ell},$$

$$\sum_{k=0}^{n} \mu_i q_k(i) q_\ell(i) = v \mu_k \delta_{k\ell}.$$

also,

$$\mu_j p_i(j) = v_i q_j(i), \quad i, j = 0, \ldots, n.$$

In matrix notation, these are,

$$P^T \Delta_\mu P = v \Delta_v,$$

$$Q^T \Delta_v Q = v \Delta_\mu,$$

where $\Delta_v = \text{diag}\{v_0, v_1, \ldots, v_n\}, \qquad \Delta_\mu = \text{diag}\{\mu_0, \mu_1, \ldots, \mu_n\}.$

Proof of Theorem:

The eigenvalues of $D_i D_\ell$ are $p_i(k)p_\ell(k)$ with multiplicities $\grave{\imath}_k$. This implies that,

$$vv_i\delta_{i\ell} = \mathrm{trace} D_i D_\ell = \sum_{k=o}^{n} \mu_i p_i(k)p_\ell(k),$$

which proves equation $\sum_{k=o}^{n} \mu_i p_i(k)p_\ell(k) = vv_i\delta_{i\ell}$, and equation $P^T\Delta_\mu P = v\Delta_v$,

$$Q = vP^{-1} = \Delta_v^{-1}P^T\Delta_\mu,$$

which gives equations $\sum_{k=o}^{n} \mu_i q_k(i)q_\ell(i) = v\mu_k\delta_{k\ell}$,

$$\mu_j p_i(j) = v_i q_j(i), \quad i,j = o,\ldots,n$$

and

$$Q^T\Delta_v Q = v\Delta_\mu.$$

There is an analogy between extensions of association schemes and extensions of finite fields. The cases we are most interested in are those where the extended schemes are defined on the n-th Cartesian power $X = \mathcal{F}^n$ of a set $\mathcal{F}$ on which a basic association scheme $(\mathcal{F},K)$ is defined. A first association scheme defined on is called the $(\mathcal{F},K)$. n-th Kronecker power $(\mathcal{F},K)_\otimes^n$ of $(\mathcal{F},K)$. Next the extension is defined on the same set $X = \mathcal{F}^n$ by gathering classes of $(\mathcal{F},K)_\otimes^n$. The Kronecker power corresponds to the polynomial ring $F[X]$ first defined on a field $\mathbb{F}$, while the extension scheme corresponds to the extension field obtained as a quotient. An example of such an extended scheme is the Hamming scheme.

Association schemes may be merged, but merging them leads to non-symmetric association schemes, whereas all usual codes are subgroups in symmetric Abelian schemes.

BUEKENHOUT GEOMETRY

A geometry, or Buekenhout geometry, then, has the following ingredients: a set X of varieties, a symmetric incidence relation I on X, a finite set Δ of types, and a type map $\tau: X \to \Delta$. We require the following axiom:

- (B1) Two varieties of the same type are incident if and only if they are equal.

In other words, a geometry is a multipartite graph, where we have names for the multipartite blocks ("types") of the graph. We mostly use familiar geometric language for incidence; but sometimes, graph-theoretic terms like diameter and girth will be useful. But one graph-theoretic concept is vital; a geometry is connected if the graph of varieties and incidence is connected.

The rank of a geometry is the number of types. A flag is a set of pairwise incident varieties. It follows from (B1) that the members of a flag have different types. A geometry satisfies the transversality condition if the following strengthening of (B1) holds:

- (B2) (a) Every flag is contained in a maximal flag.
- (b) Every maximal flag contains one variety of each type.

All geometries here will satisfy transversality.

Let F be a flag in a geometry G. The residue $G_F = R(F)$ of F is defined as follows: the set of varieties is,

$$X = \{x \in X \setminus F : xIy \text{ for all } y \in F\};$$

the set of types is $\Delta_F = \Delta \setminus \tau(F)$; and incidence and the type map are the restrictions of those in G. It satisfies (B1) (resp. (B2)) if G does. The type of a flag or residue is its image under the type map, and the cotype is the complement of the type in Δ; so the type of GF is the cotype of F. The rank and corank are the cardinalities of the type and cotype.

A transversal geometry is called thick (resp. firm thin) if every flag of corank 1 is contained in at least three (resp. at least two, exactly two) maximal flags.

A property holds residually in a geometry if it holds in every residue of rank at least 2. (Residues of rank 1 are sets without structure.) In particular, all geometries of interest are residually connected; in effect, we assume residual connectedness as an axiom:

- (B3) All residues of rank at least 2 are connected.

Proposition: Let G be a residually connected transversal geometry, and let x and y be varieties of X, and i and j distinct types. Then there is a path from x to y in which all varieties except possibly x and y have type i or j.

Proof: The proof is by induction on the rank. For rank 2, residual connectedness is just connectedness, and the result holds by definition. So assume the result for all geometries of smaller rank than G.

We show first that a two-step path whose middle vertex is not of type i or j can be replaced by a path of the type required. So let xzy be a path of length 2. Then x and y lie in the residue of z; so the assertion follows from the inductive hypothesis.

Now this construction reduces by one the number of interior vertices not of type i or j on a path with specified endpoints. Repeating it as often as necessary gives the result.

Let Δ be a finite set. Assume that, for any distinct i j $\in \Delta$, a class $\mathcal{G}_{ij}$ of geometries of rank 2 is given, whose two types of varieties are called "points" and "blocks". Suppose that the geometries in $\mathcal{G}_{ij}$ are the duals of those in $\mathcal{G}_{ij}$. The set Δ equipped with these collections of geometries is called a diagram. It is represented pictorially by taking a "node" for each element of Δ, with an "edge" between each pair of nodes, the edge from i to j being adorned or labelled with some symbol for the class ${}_{ij}$.

A geometry G belongs to the diagram $\left(\Delta,\left(\mathcal{G}_{ij} : i, j \in \Delta\right)\right)$ if Δ is the set of types of G and, for all distinct $i\,j \in \Delta$, and all residues GF in G with rank 2 and type $\{i, j\}$, GF is isomorphic to a member of $\mathcal{G}_{ij}$ (where we take points and blocks in GF to be varieties of types i and j respectively).

In order to illustrate this idea, we need to define some classes of rank 2 geometries to use in diagrams. Some of these we have met already; but the most important is the most trivial: A digon is a rank 2 geometry (having at least two points and at least two blocks) in which any point and block are incident; in other words, a complete bipartite graph containing a cycle. By abuse of notation, the "labelled edge" used to represent digons is the absence of an edge! This is done in part because most of the rank 2 residues of our geometries will be digons, and this convention leads to uncluttered pictorial representations of diagrams.

A partial linear space is a rank 2 geometry in which two points lie on at most one line (and dually, two lines meet in at most one point). It is represented by an edge with the label Π thus:

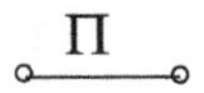

We already met the concepts linear space and generalised projective plane: they are partial linear spaces in which the first, resp. both, occurrences of "at most" are replaced by "exactly". They are represented by edges with label L and without any label, respectively. (Conveniently, the labels for the self-dual concepts of "partial linear space" and "generalised projective plane" coincide with their mirror-images, while the label for "linear space" does not.) Note that a projective plane is a thick generalised projective plane. Another specialisation of linear spaces, a "circle" or "complete graph", has all lines of cardinality 2; it is denoted by an edge with label c.

Proposition: A projective geometry of dimension n has the diagram:

Proof: Transversality and residual connectivity are straightforward to check. We verify the rank 2 residues. Take the types to be the dimensions $0, 1, \ldots n-1$, and let F be a flag of cotype $\{i, j\}$, where $i < j$.

Case: $j = i + 1$. Then F has the form,

$$U_0 < U_1 < \ldots < U_{i-1} < U_{i+2} < \ldots < U_{n-1}.$$

Its residue consists of all subspaces of dimension i or $i + 1$ between U_{i-1} and U_{i+2}; this is clearly the projective plane based on the rank 3 vector space U_{i+2} / U_{i-1}.

Case: $j > i+1$. Now the flag F looks like,

$$U_0 < \ldots < U_{i-1} < U_{i+1} < \ldots < U_{j-1} < U_{j+1} < \ldots < U_{n-1}.$$

Its residue consists of all subspaces lying either between U_{i-1} and U_{i+1} or between U_{j-1} and U_{i+1} or between U_{j-1} and U_{j+1} Any subspace X of the first type is incident with any subspace Y of the second, since $X < U_{i+1} \leq U_{j-1} < Y$. So the residue is a digon.

In diagrams, it is convenient to label the nodes with the corresponding elements of Δ .For example, in the case of a projective geometry of dimension n, we take the labels to be the dimensions of varieties represented by the nodes, thus:

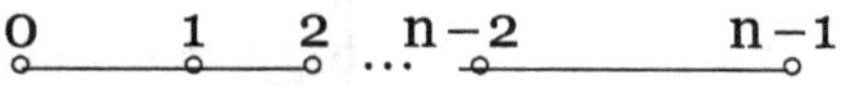

This reserves the space below the nodes for another use, as follows.

A transversal geometry is said to have orders, or parameters, if there are numbers S_i (for $i \in \Delta$) with the property that any flag of cotype i is contained in exactly $S_i + 1$ maximal flags. If so, these numbers S_i are the orders (or parameters). Now, if G is a geometry with orders, then G is thick/firm/thin respectively if and only if all orders are $>1/\geq 1/=1$ respectively. We will write the orders beneath the nodes, where appropriate. Note that a projective plane of order n (as defined earlier) has orders n n (in the present terminology). Thus, the geometry $PG(n,q)$ has diagram.

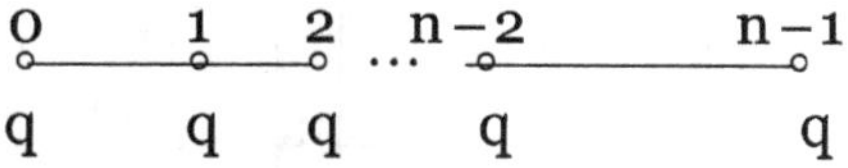

Proposition: Let the diagram Δ be the disjoint union of Δ_1 and Δ_2 with no edges between these sets. Then a variety with type in and one with type in Δ_2 are incident.

Proof: We use induction on the rank. For rank $2, \Delta$ is the diagram of a digon, and the result is true by definition. So assume that $|\Delta| > 2$, and (without loss of generality) that $|\Delta_1| > 1$.

Let X_i be the set of varieties with type in Δ_1, for $i = 1,2$. By the inductive hypothesis, if $i = 1,2$. By the inductive hypothesis, if $x, y \in X_1$ with xly then $R(x) \cap X_2 = R(y) \cap X_2$. (Considering $R(x)$, we see that every variety in $R(x) \cap X_2$ is incident with y so the left-hand set is contained in the right-hand set. Reversing the roles of x and y establishes the result.) Now by connectedness $R(x) \cap X_2$ is independent of $x \in X_1$. But this set must be X_2 since every variety in X_2 is incident with some variety in X_1.

A diagram is linear if the "non-digon" edges form a simple path, as in the diagram for projective spaces in proposition above.

Suppose that one particular type in a geometry is selected, and varieties of that type are called points. Then the shadow, or point-shadow, of a variety x is the set Sh (x) of varieties incident with x. Sometimes we write Sho (x) where o is the type of a point. In a geometry with a linear diagram, the convention is that points are varieties of the left-most type.

Corollary: In a linear diagram, if xly, and the type of y is further to the right than that of x, then $sh(x) \subseteq sh(y)$.

Proof: $R(x)$ has disconnected diagram, with points and the type of y in different components; so, by proposition above, every point in $R(x)$ is incident with y.

Example:

Construct a geometry which is connected but not residually connected.

Show that, if G has any of the following properties, then so does any residue of G of rank at least 2: residually connected, transversal, thick, firm, thin.

Show that any generalised projective geometry belongs to the diagram.

A chamber of a transversal geometry G is a maximal flag. Let F be the set of chambers of the geometry G. Form a graph with vertex set F by joining two chambers which coincide in all but one variety. G is said to be chamberconnected if this graph is connected. Prove that a residually connected geometry is chamber-connected, and a chamber-connected geometry is connected.

Consider the 3-dimensional affine space $AG(3,F)$ over the field F. Take three types of varieties: points (type 0), lines (type 1), and parallel classes of planes (type 2). Incidence between points and lines is as usual; a line L and a parallel class C of planes are incident if L lies in some plane of C; and any variety of type 0 is incident with any variety of type 2. Show that this geometry is chamber-connected but not residually connected.

LetV be a six-dimensional vectorspace over a field F, with a basis $\{e_1,e_2,e_3,f_1,f_2,f_3\}$. Let G be the additive group of V, and let H_1,H_2,H_3 be the additive groups of the three subspaces $\langle e_3,e_2,f_1\rangle$, $\langle e_3,e_1,f_2\rangle$, and $\langle e_1,e_2,f_3\rangle$. Form the coset geometry $G(G,(H_1,H_2,H_3))$: its vaarieties of type i are the cosets of H_i in G, and two varieties are incident if and only if the corresponding cosets have non-empty intersection. Show that this geometry is connected but not chamber-connected.

Some Special Diagrams

We first consider geometries with linear diagram in which all strokes are linear spaces; then we specialise some or all of these linear spaces to projective or affine planes. We will see that the axiomatisations of projective and affine spaces can be expressed very simply in this formalism.

Theorem: Let G be a geometry with diagram,

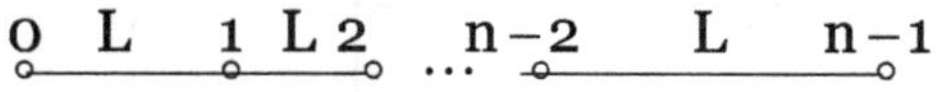

Let varieties of type 0 and 1 be points and lines.

- The points and shadows of lines form a linear space L .
- The shadow of any variety is a subspace of L .
- $Sh_0(x) \subseteq Sh_0(y)$ if and only if x is incident with y.
- If x is a variety and p a point not incident with x, then there is a unique variety y incident with x and p such that $\tau(y) = \tau(x) + 1$.

Proof: (a) We show that two points lie on at least one line by induction on the rank. There is a path between any two points using only points and lines; so it suffices to show that any such path of length greater than 2 can be shortened. So assume pILIqIMlr, where p,q,r are points and L,M lines. By the induction hypothesis, the POINTs L and M of R(q) lie in a LINE Π , a plane of G incident with L and M. By Corollary, p and q are incident with Π . Since Π is a linear space, there is a line through p and q. (The convention of using capitals for varieties in R (q) is used here).

Now suppose that two lines L and M contain the two points p and q. Considering R(p), we find a plane Π incident with L and M and hence with p and q. But Π is a linear space, so $L = M$.

Let y be any variety, and $p, q \in Sh_0(y)$. Since points and lines incident with y form a linear space by (a), there is a line incident with p q and y. This must be the unique line incident with p and q; and, by Corollary 5.4, all its points are incident with y and so are in $Sh_0(y)$.

The reverse implication is Corollary. So suppose that $Sh_0(x) \subseteq Sh_0(y)$. Take $p \in Sh_0(x)$. Then, in $R(p)$, we have $Sh_1(x) \subseteq Sh_1(y)$ (since these shadows are linear subspaces), and so xly by induction. (The base case of the induction, where x is a line, is covered by (b).)

This is clear if x is a point. Otherwise, choose $q \in Sh_0(x)$, and apply induction in $R(q)$ (replacing p by the line pq).

Theorem: A geometry with diagram,

is a generalised projective space (of finite dimension).

Proof: By Theorem, a potential Veblen configuration lies in a plane; since planes are projective, Veblen's axiom holds. It remains to show that every linear subspace is the shadow of some variety; this follows easily by induction.

Theorem: A geometry with diagram,

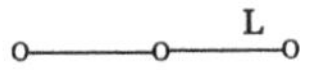

consists of the points, lines and planes of a (possibly infinite-dimensional) generalised projective space.

Proof: Veblen's axiom is verified as in Theorem 5.6. It is clear that every point, line or plane corresponds to a variety.

By the argument for Theorem, we have all the points, lines and planes, and some higher-dimensional varieties, of a generalised projective space. Examples arise by taking all the flats of dimension at most $r-1$, where r is the rank. However, there are other examples. A simple case, with $r=4$, can be constructed as follows.

Let P be a projective space of countable dimension over a finite field F. Enumerate the 3-dimensional and 4-dimensional subspaces in lists $T_0, T_1, \ldots$ and $F_0, F_1, \ldots$. Now construct a set F of 4-dimensional subspaces in stages as follows. At the nth stage, if T_n is already contained in a member of F, do nothing. Otherwise, of the infinitely many subspaces F_j which contain T_n, only finitely many are excluded because they contain any T_m with $m<n$; let F_i be the one with smallest index which is not excluded, and adjoin it to F. At the conclusion, any 3-dimensional subspace is contained in a unique member of F. Then the points, lines, planes, and subspaces in F form a geometry with the diagram.

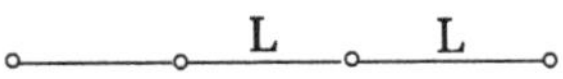

where the first L denotes the points and lines in 3-dimensional projective space over F.

The label Af on a stroke will denote the class of affine planes.

Theorem: A geometry with diagram,

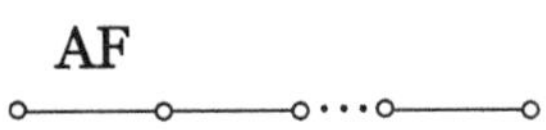

is an affine space of finite dimension.

Proof: It is a linear space whose planes are affine. We must show that parallelism is transitive. So suppose that $L_1 \| L_2 \| L_3 \|$, but $L_1 \nparallel L_3$. Then all three lines lie in a subspace of dimension 3; so it is enough to deduce a contradiction in the case of geometries of rank 3. Note that, for a geometry with diagram 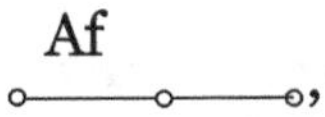

, two planes which have a common point must meet in a line.

Let Π_1 be the plane through L_1 and L_2, and Π_2 the plane through p and L_3, where p is a point of L_1. Then Π_1 and Π_2 both contain p, so they meet in a line $M \neq L_1$. Then M is not parallel to L_2, so meets it in a point q, But then Π_2 contains L_3 and q, hence L_2, and so is equal to Π_1, a contradiction.

Theorem: A geometry with diagram,

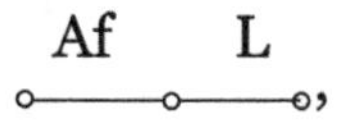

in which some line has more than three points, consists of the points, lines and planes of a (possibly infinite-dimensional) affine space.

Example: Consider a geometry of rank n with diagram,

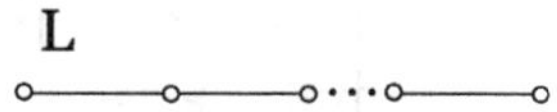

in which all lines have the same finite cardinality k, and all the projective planes have the same finite order q.

- If $n \geq 4$, prove that the geometry is either projective $(q-k-1)$ or affine $(q=k)$.
- If $n=3$, prove that $q=k-1, k, k^2$ or $k(k^2+1)$.

Construct an infinite "free-like" geometry with diagram:

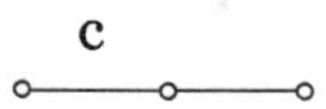

(Ensure that three points lie in a unique plane, while two planes meet in two points).

Show that an inversive plane belongs to the diagram:

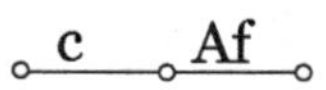

What are the varieties?

Show how to construct a geometry with diagram:

(n nodes) from an ovoid in $PG(n,F)$.

References

- Macdonald, I. G. Symmetric functions and Hall polynomials. Second edition. Oxford Mathematical Monographs. Oxford Science Publications. The Clarendon Press, Oxford University Press, New York, 1995. x+475 pp. ISBN 0-19-853489-2 MR1354144
- Algebraic-combinatorics, research-areas, research-combinatorics-and-optimization, combinatorics-and-optimization: uwaterloo.ca, Retrieved 14 January, 2020
- Stembridge, John R. (2002), "A concise proof of the Littlewood–Richardson rule" (PDF), Electronic Journal of Combinatorics, 9 (1): Note 5, 4 pp. (electronic), ISSN 1077-8926, MR 1912814
- Coherent-Algebras-236013383: researchgate.net, Retrieved 23 March, 2020

Diverse Areas of Combinatorics

CHAPTER 7

There are various areas that fall under the domain of combinatorics such as analytic combinatorics, infinitary combinatorics, arithmetic combinatorics, etc. This chapter has been carefully written to provide an easy understanding of these diverse areas of combinatorics.

ANALYTIC COMBINATORICS

Analytic combinatorics is the branch of combinatorics that analyzes families of combinatorial objects using their generating functions. Those are series which coefficients contain the combinatorial information on the objects. This field has many qualities that make it great for the analysis of random graphs:

- From the generating functions, many combinatorial information can be extracted, such as precise asymptotics (with as many error terms as wanted), and moments and limit laws of parameters,
- The generating functions can be combined to represent new interesting objects,
- The method is robust: A small perturbation of the model requires often only a small adjustment in the generating functions.

A multiset is a collection of objects, without order, where repetitions are allowed. A set is then a multiset without repetitions, and a sequence, or list, or tuple, is an ordered multiset. We denote multisets and sets by the bracket notation {2, 3, 7}, and sequences by the parenthesis notation (3, 2, 7). The nth coefficient of the generating function,

$$F(Z) = \sum_{n \geq o} f_n^{z^n}$$

is denoted by fn = $[z^n]f(z)$. The derivative of the function f is denoted by ∂f or f'.

Technical Lemmas

Our primary objective is to derive exact expressions for the number of graphs that satisfy some properties. Those numbers will be expressed as the coefficients of generating functions, characterized by various relations. Analytic combinatorics has developed many "black box" theorems that can be applied to obtain the asymptotics of generating function coefficients. The choice of the theorem depends of the form of the generating function. During the first reading, we suggest not to spend too much time on the technical conditions of those theorems, but rather to recognize the main features.

Theorem: (Singularity analysis). We consider a series f(z) of positive radius of convergence ρ, analytic on the set,

$$\Delta = \{z \,|\, |z| < R,\ z \neq \rho,\ |\arg(z-\rho)| > \phi\}$$

for some values R > ρ and $0 < \phi < \frac{\pi}{2}$. If,

$$f(z) \sim (1 - z/\rho)^{-\alpha} \text{ as } z \to \rho \text{ while } z \in \Delta,$$

for some α/∈ {0, −1, −2,...}, then,

$$[z^2] f(z) \sim \frac{\rho^{-n} n^{\alpha-1}}{\Gamma(\alpha)}.$$

The following theorem analyses the singularity of generating functions characterized implicitly. This is in particular the case for trees.

Theorem: (Implicit functions). Consider a function ϕ(u) analytic at u = 0, with nonnegative coefficients, ϕ(0) ≠ 0, and that is not of the form $\phi(u) = \phi_0 + \phi_1 u$. Furthermore, assume that the equation,

$$\phi(\tau) - \tau\phi'(\tau) = 0$$

admits a real positive solution, smaller than the radius of convergence of ϕ. Then the function y(z) defined implicitly by the relation,

$$y(z) = z\phi(y(z))$$

has radius of convergence $\rho = \frac{\tau}{\phi(\tau)}$ and is analytic on a set Δ of the form given in theorem. The Laplace method is a classic analytic technique.

Theorem: (Laplace method). We consider a neighborhood C of the origin in $\mathbb{R}^d$, and two analytic functions A and φ from C to $\mathbb{C}$. Suppose that the real part of φ(x) – φ(0) is strictly positive on C except at the origin, and that its Hessian matrix H is nonsingular there.

If A does not vanish at the origin, then,

$$\int_{x\in C} A(x)e^{n\phi(x)dx} \sim \frac{A(0)e^{n\phi(0)}}{\sqrt{\det(H)}}\left(\frac{2\pi}{n}\right)^{d/2}.$$

The Cauchy integral transforms a coefficient extraction into a complex integral on a small loop around the origin,

$$[z^n]f(z) = \frac{1}{2i\pi}\oint \frac{f(z)}{z^n}\frac{dz}{z} \overset{z=\zeta e^{i\theta}}{=} \frac{1}{2\pi}\int_{-\pi}^{\pi} \frac{f(\zeta e^{i\theta})}{\zeta^n e^{in\theta}}d\theta.$$

A corollary of the Laplace method is then the large powers theorem.

Theorem: (Large powers theorem). We consider integers n, N, such that N/n has a positive limit λ. Let $B(z) = \Sigma_{n\geq 0} b_n z^n$ be a series with nonnegative coefficients and radius of convergence ρ_B, that satisfies $\gcd\{i - j \mid bi \neq 0, b_j \neq 0\} = 1$ (thus B(z) is not a function of the form $z^r C(z^p)$ for some integer p ≥ 2 and some function C analytic at 0). We introduce the function $L(z) = \frac{zB'(z)}{B(z)}$, and assume that the equation L(z) = λ has a positive solution ζ < ρB (which is then unique), and that $L'(\zeta) \neq 0$. Let A(z) be a generating function with radius of convergence greater than ζ, that does not vanish at ζ. Then,

$$[z^N]A(z)B(z)^n \sim \frac{A(\zeta)}{\sqrt{2\pi n\zeta L'(\zeta)}}\frac{B(\zeta)^n}{\zeta^N}.$$

The value ζ from the large powers theorem is called the saddle-point.

Labels

A labelled object is a set of labelled atoms, with some structure on it. In a tree, for example, the atoms are the vertices. The size of an object a, denoted by |a|, is then its number of atoms. The labels on the atoms are distinct integers. Given an object of size n, there exists a unique way to relabel its atoms in {1, 2,..., n}, so that the relative order of the atoms stays the same. When counting labelled objects, we thus assume without lost of generality that the labels are consecutive integers starting at 1. For example, a permutation of size n can be represented as a sequence of n distinct integers from {1, 2,..., n}. Therefore, permutations are labelled combinatorial objects.

We use an exponential generating function to represent a labelled combinatorial family $\mathcal{F}$.

$$F(z) = \sum_{a\in\mathcal{F}} \frac{z^{|a|}}{|a|!} = \sum_{n\geq 0} f_n \frac{z^n}{n!},$$

where f_n denotes the number of objects of size n in $\mathcal{F}$. The reason of this convention is that many natural combinatorial constructions on labelled families translate well into exponential generating function operations:

- The generating function of the disjoint union of two families is the sum of their generating function:

  $C = A \uplus \mathcal{B}$, implies $C(z) = A(z) + B(z)$.

- The relabelled Cartesian product C of two labelled families A and B is defined as the pairs in A × B, where the two objects are relabelled to ensure that the atoms have distinct labels. For example, the pair of permutations ((1, 3, 2),(2, 1)) is not a proper labelled object, since the atoms 1 and 2 appears twice. The ten corresponding relabelled pairs of permutations are:

  ((1, 3, 2),(5, 4)), ((1, 4, 2),(5, 3)), ((1, 4, 3),(5, 2)), ((1, 5, 2),(4, 3)), ((1, 5, 3),(4, 2)), ((1, 5, 4),(3, 2)), ((2, 4, 3),(5, 1)), ((2, 5, 3),(4, 1)), ((2, 5, 4),(3, 1)), ((3, 5, 4),(2, 1)).

Observe that the relative orders of the atoms are preserved. The generating function of the relabelled Cartesian product of two labelled families is,

$$C = A * B, \text{ implies } C(z) = A(z)B(z).$$

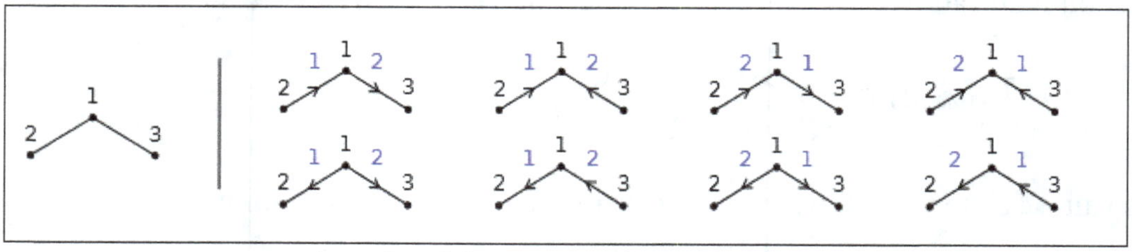

A graph and the corresponding multigraphs.

The generating function of sequences of labelled objects from A is 1/(1 – A(z)) (again, a relabelling occurs). Indeed, the combinatorial family equal to this sequence is $\cup_{n\geq 0} A^n$, which has generating function $\sum_{n\geq 0} A(z)^n = \frac{1}{1-A(z)}$, .

The generating function of sets of labelled objects from A is exp(A(z)). Indeed, a set of n objects from A is a sequence, considered up to any of the n! permutations, so the family of sets of n objects from A has generating function $\frac{A(z)^n}{n!}$, and the union over n leads to the exponential.

The generating function of oriented cycles of labelled objects from A is $\log\left(\frac{1}{1-A(z)}\right)$. A cycle of n objects from A is a sequence, considered up to any of the n circular permutations, so the generating function of those cycles is $\frac{A(z)^n}{n}$, and the sum over n is the logarithm.

For the analysis of graphs, we will use two different kinds of atoms: the vertices and the edges.

Models: Graphs and Multigraphs

We define a multigraph G = (V, E) as a labelled set V of vertices, and a labelled multiset E of edges, where each edge is an oriented pair of vertices. An edge e is then a triplet (u, v, ℓ), where u and v are the vertices linked by e, and ℓ` is the label of e.

The combinatorics and graph theory communities usually work on graphs instead of multigraphs. The difference is that in a graph, the edges are unlabelled and unoriented. Furthermore, loops (an edge linking a vertex to itself) and multiple edges (set of edges linking the same two vertices) are forbidden. However, multigraphs turn out to be better suited for generating function manipulations than graphs. All the results presented can be derived for graphs as well. Examples of graphs and multigraphs are displayed in figure. The number of vertices of a multigraph G is denoted by n(G), and its number of edges by m(G). We also define its excess as k(G) = m(G) – n(G).

Since multigraphs have labelled vertices and labelled edges, we use for them generating functions exponential with respect to both z and w. Furthermore, because their edges are oriented, we introduce a weight 1/2 on them. The generating function of a multigraph family $\mathcal{F}$ is then defined as:

$$F(z,w)=\sum_{G\in\mathcal{F}}\frac{w^{m(G)}}{2^{m(G)}m(G)!}\frac{z^n(G)}{n(G)!}.$$

Trees and Unicycles

Recall that the excess of a graph is the difference between its number of edges and vertices. Trees have excess −1, which is the minimum possible excess for a connected graph. A unicycle is a connected multigraph of excess 0.

Theorem: The generating functions of rooted trees, trees, and unicycles are characterized by the relations,

$$T(z)=zeT^{(z)}$$

$$U(z)=T(z)-T(z)^2/2$$

$$V(z)=\frac{1}{2}\log\left(\frac{1}{1-T(z)}\right)$$

Proof: A rooted tree is a vertex (the root) and a set of sons, which are themselves rooted trees, so,

$$T(z) = ze^{T(z)}.$$

Each tree of size n correspond to n rooted trees (number of possible choices for the root), so,

$$zU'(z) = \sum_{n\geq 0} nu_n \frac{z^n}{n!} = T(z),$$

and we can check that T(z)–T(z) ²/2 is the unique solution of this differential equation. Any unicycle can be uniquely decomposed as a non-oriented cycle, where each vertex is replaced by a rooted tree so,

$$V(z) = \frac{1}{2}\log\left(\frac{1}{1-T(z)}\right)$$

Asymptotics expressions for the number of trees and unicycles with n vertices can be extracted using singularity analysis. We can also prove that random multigraphs with a small number of edges typically contain only trees and unicycles, a result first derived by Erd″os and R′enyi. To do so, we compare the number of such multigraphs to the total number of multigraphs. Recall that the excess of a multigraph is the difference between its number of edges and vertices.

Theorem: When m/n tends toward a constant smaller than 1/2, almost all multigraphs with n vertices and m edges contain only trees and unicycles.

Proof: Trees have excess −1, and unicycles excess 0. Therefore, a multigraph of excess k = m − n (which is negative when m/n < 1/2) that contains only trees and unicycles is a set of −k trees and a set of unicycles,

$$\frac{U(z)^{-k}}{(-k)!} e^{V(z)}.$$

So the number of such multigraphs with n vertices and m edges is,

$$n!2^m m!\left[z^n\right]\frac{U(z)^{n-m}}{(n-m)!} e^{V(z)}.$$

We apply the large powers theorem to extract the asymptotics of this expression,

$$n!2^m m!\left[z^n\right]\frac{U(z)^{n-m}}{(n-m)!} e^{V(z)} \sim \frac{n!2^m m!}{(n-m)!} \frac{U(\zeta)^{n-m} e^{V(\zeta)}}{\zeta^n \sqrt{2\pi(-k)\zeta\phi'(\zeta)}},$$

where $\phi(z)=\dfrac{zU'(z)}{U(z)}$ and ζ is the unique positive solution of $\phi(\zeta)=\dfrac{n}{n-m}$. After some computations, we find,

$$n!2^m m![z^n]\,\frac{U(z)^{n-m}}{(n-m)!}e^{V(z)} \sim n^{2m},$$

which is the total number of multigraphs with n vertices and m edges. Therefore, when m/n has a limit smaller than 1/2, almost all multigraphs with n vertices and m edges contain only trees and unicycles.

Connected Multigraphs with Fixed Excess

Lemma: The number of kernels of a given excess is finite: a kernel of excess k contains at most 2k vertices and 3k edges. Those bounds are reached by cubic multigraphs, i.e. multigraphs where all vertices have degree exactly 3.

Proof: We consider any kernel with n vertices, m edges, and excess k = m − n. The sum of the degrees of all vertices is equal to twice the number of edges, and each vertex has degree at least 3, so,

$$2m = \sum_{\text{vertex } \upsilon} \deg(\upsilon) \geq 3n,$$

which implies $n \leq 2k$ and $m \leq 3k$. Those bounds are reached when deg(v) = 3 for each vertex v.

We derived the generating functions of connected multigraphs with excess −1 (trees) and 0 (unicycles). We now consider connected multigraphs with positive excess.

Theorem: For any $k \geq 1$, there exists a computable polynomial $Q_k(T)$ such that the generating function of connected multigraphs of excess k is,

$$CMG_\kappa(z)=\frac{Q_\kappa(T(z))}{(1 - T(z))^{3\kappa}}.$$

Proof: Let us define a path of trees as a sequence,

$$(\text{edge}, \text{rooted tree}, \text{edge}, \text{rooted tree}, \ldots, \text{edge})$$

where the vertices are labelled, and the edges are labelled and oriented. Each edge links the roots of the two neighbor trees in the sequence, except the first and last edges. The generating function of path of trees is,

$$\frac{1}{1-T(z)}.$$

Observe that a path of trees contains one more edge than its number of vertices. Any connected multigraph with positive excess can be uniquely decomposed as a connected kernel where:

- Vertices are replaced by rooted trees.
- Edges are replaced by paths of trees.

By construction, the kernel has the same excess as the multigraph. The generating function of connected kernels of excess k is a multinomial $CK_k(z,w)$ of power $3k$ in w. Therefore,

$$CMG_k(z) = CK\left(T(z), \frac{1}{1-T(z)}\right) = \frac{Q_k(T(z))}{(1-T(z))^{3k}}.$$

The asymptotics of connected multigraphs with n vertices and excess k is then derived by application of a singularity analysis.

The generating function of multigraphs of excess k that contain trees and unicycles and exactly $C\ell$ components of excess ℓ for all $1 \le \ell \le L$ is,

$$\frac{U(z)^{-k+K}}{(-k+K)!} e^{V(z)} \frac{\prod_{\ell=1}^{L} Q_\ell (T(z))^{c\ell}}{(1-T(z))^{3K}},$$

where $K = \Sigma_{\ell=1}^{L} \ell_{c\ell}$. The limit probability for a random graph with n vertices and m edges to contain exactly c` components of excess ℓ for all $1 \le \ell \le L$ is non-zero only when $m = \frac{n}{2}\left(1+O\left(n^{-1}/3\right)\right)$, and they computed this limit probability in that case.

Using a probabilistic approach, Erd″os and R´enyi proved that when m/n has a limit greater than 1/2, a typical random graph with n vertices and m edges contains only trees, unicycles, and a unique component of positive excess, called the giant component. Analyzing the statistics of this giant component using analytic combinatorics is an open problem.

A multigraph and its representation as a set of vertices with labelled half-edges.

Multigraphs with Degree Constraints

The goal of this section is the enumeration of multigraphs with n vertices, m edges, and where each vertex has its degree in a given set D. We denote by,

$$Set_D(z) = \sum_{d \in D} \frac{z^d}{d!}$$

The exponential generating function of this set, and assume that:

- D contains at least 2 elements.
- $\gcd\{d1 - d_2 \mid d_1, d_2 \in D\} = 1$.

The first assumption discards the enumeration of regular multigraphs, where all vertices have the same degree, and that can be analyzed separately. The second assumption just simplifies the analysis, and the general case has been treated by de Panafieu and Ramos.

Theorem : The number of multigraphs with n vertices, m edges, and all vertices having their degree in the set D is,

$$(2m)!\left[x^{2m}\right]\mathrm{Set}_D(x)^n.$$

Proof: Let us consider a multigraph G, and cut each edge into two labelled half-edges. Specifically, an edge labelled ℓ and oriented from the vertex u to the vertex v is replaced by a half-edge labelled $2\ell - 1$ and attached to u, and a half-edge labelled 2ℓ and attached to v. This transforms the multigraph G into a set of vertices, to each of which is attached a set of labelled vertices. The size of each of those sets is the degree of the vertex, and the total number of half-edges is twice the initial number of edges. Therefore, the number of graphs with n vertices, m edges, and having all their degrees in D is,

$$(2m)!\left[x^{2m}\right]\mathrm{Set}_D(x)^n.$$

Connected Multigraphs with Large Excess

The generating function of all multigraphs is,

$$MG(z,w) = \sum_{n \geq 0} e^{n^2 w/2} \frac{z^n}{n!},$$

because when ordering the edges according to their labels and orientations, a multigraph with n vertices and m edges becomes a sequence of 2m vertices in {1, 2,..., n}, so there are n^{2m} such multigraphs. Since a multigraph is a set of connected multigraphs, the generating function of connected multigraphs CMG(z, w) satisfies the relation,

$$MG(z, w) = e^{CMG(z,w)}.$$

Taking the logarithm, we obtain the classic closed form for the generating function of connected multigraphs,

$$CMG(z,w) = \log\left(\sum_{n \geq 0} e^{n^2 w/2} \frac{z^n}{n!}\right)$$

Observe that the argument of the logarithm is a series with a zero radius of convergence. Therefore, we cannot use any analytic property of the logarithm, and the only way to extract the asymptotics seems to be to expand it as a series,

$$\mathrm{CMG}(z,w)=\sum_{q\geq 1}\frac{(-1)^{q+1}}{q}\left(\sum_{n\geq 0} e^{n^2 w/2}\frac{z^n}{n!}\right)^q.$$

This expression was the starting point of the analysis of Flajolet et al., who worked on connected graphs with fixed excess. If we extract the coefficient $n!2^m m![z^n w^m]$, we obtain an exact expression for the number of connected multigraphs with n vertices and m edges,

$$\mathrm{CMG}_{n,m}=\sum_{q-1}^{n}\frac{(-1)^{q+1}}{q}\sum_{\substack{n_1+\cdots+n_q=n\\ \forall j,\ n_j\geq 1}}\binom{n}{n_{1,\ldots,}n_q}\left(n_1^2+\cdots+n_q^2\right)^m$$

However, as already observed by those authors, it is difficult to extract the asymptotics, because of "magical" cancellations in the coefficients. In particular, the dominant contribution to the sum does not come from the first value q = 1, because the summand is then the number of (non-empty) multigraphs with n vertices and m edges. Those multigraphs are indeed typically not connected, as they contain many trees and unicycles. Instead of working on this expression using complicated analysis, we will derive a different (although similar) expression, better suited for asymptotics analysis. The main idea, already applied by Pittel and Wormald, is to consider the family MG>o of multigraphs without trees and unicycles. We call them positive multigraphs, since all their components have a positive excess. Let $\mathrm{CMG}^{>o}$ denote the set of connected multigraphs with positive excess. A set of connected multigraphs with positive excess is either empty, or is a positive multigraph, so,

$$e^{\mathrm{CMG}>o}(z,w)=1+\mathrm{MG}^{>o}(z,w),\quad \text{which implies}\quad \mathrm{CMG}^{>o}(z,w)=\log\left(1+\mathrm{MG}^{>o}(z,w)\right)$$

Working with the excess instead of the number of edges, and denoting by $\mathrm{CMG}_k(z) = [y^k$ $]\ \mathrm{CMG}(z/y, y)$ the generating function of connected graphs of excess k, we obtain the following expression:

$$\mathrm{CMG}_\kappa(z)=\left[y^\kappa\right]\log\left(1+\sum_{\ell>o}\mathrm{MG}_\ell^{>o}(z)y^\ell\right)=\sum_{q\geq 1}\frac{(-1)^q}{q}\sum_{\substack{\kappa_1+\cdots+k_q=\kappa\\ \forall j,\kappa_j\geq 1}}\prod_{j=1}^{q}\mathrm{MG}_{\kappa_j}^{>o}(z)$$

However, the dominant contribution to the sum will be easy to locate: it comes from the term q =1:

$$n!\left[z^n\right]\mathrm{CMG}_\kappa(z)\sim n!\left[z^n\right]\mathrm{MG}_\kappa^{>o}(z)$$

Theorem: The generating function of positive multigraphs of excess k is,

$$MG_k^{>0}(z)=\frac{(2k)!}{2^k k!}\left[x^{2k}\right]\frac{e^{-V(z)}}{\left(1-T(z)\dfrac{e^x-1-x}{x^2/2}\right)^{k+1/2}}$$

Proof: A core is a multigraph of minimum degree 2., The generating function of cores is,

$$Core(z,w)=\sum_{m\geq 0}(2m)!\left[x^{2m}\right]e^{z\left(e^x-1-x\right)}\frac{w^m}{2^m m!}.$$

In this expression, after developing the exponential as a sum over n and applying the change of variable m ← k + n, we obtain,

$$Core(z,w)=\sum_{k\geq 0}\left[x^2\right]\sum_{n\geq 0}\frac{(2(k+n))!}{2^{k+n}(k+n)!}\frac{\left(zw\dfrac{e^x-1-x}{x^2}\right)}{n!}w^k.$$

The sum over n is replaced by its closed form,

$$Core(z,w)=\sum_{k\geq 0}\left[x^{2k}\right]\frac{(2k)!}{2^k k!}\frac{w^k}{\left(1-zw\dfrac{e^x-1-x}{x^2/2}\right)^{k+1/2}}$$

The generating function of multicores of excess k is then,

$$Core_k(z)=\left[y^k\right]Core(z/y,y)=\frac{(2k!)}{2^k K!}\left[x^{2k}\right]\frac{1}{\left(1-zw\dfrac{e^x-1-x}{x^2/2}\right)^{k+1/2}}$$

In a multigraph, if we remove again and again all vertices of degree 0 and 1, the trees disappear, and the rest of the multigraph is reduced to a core. Conversely, any positive multigraph with a set of unicycles can be uniquely decomposed as a core where each vertex is replaced by a rooted tree. Furthermore, the core and the multigraph have the same excess, so,

$$MG_k^{>0}(z)e^{V(z)}=Core_k(T(z)).$$

This implies,

$$MG_k^{>0}(z)e^{V(z)}=Core_k(T(z))e^{-V(z)}=\frac{(2k!)}{2^k K!}\left[x^{2k}\right]\frac{e^{-V(z)}}{\left(1-T(z)\dfrac{e^x-1-x}{x^2/2}\right)^{k+1/2}}.$$

What we gained with this new expression of the asymptotic number of connected multigraphs with n vertices and excess k = m – n,

$$n!2^{k+n}(k+n)!\left[z^n\right]CMG_k(z) \sim n!2^{k+n}(k+n)!\frac{(2k)!}{2^k k!}\left[z^n x^{2k}\right]\frac{e^{-V(z)}}{\left(1-T(z)\frac{e^{x-1-x}}{x^2/2}\right)^{k+1/2}}$$

is that the right-side expression can be analyzed using a bivariate large powers theorem. We express the coefficient extractions $[z^n x^{2k}]$ as Cauchy integrals and apply the Laplace method.

Multigraphs with Forbidden Subgraphs

We consider a connected multigraph H that is not a tree, and assume it is strictly balanced, which means that its density is greater than the density of any of its subgraphs,

$$\text{for all } G \subsetneq H, \frac{m(H)}{n(H)} > \frac{m(G)}{n(G)}.$$

We derive the limit probability for a random multigraph with n vertices and m edges to contain a copy of H as a subgraph. As usual, a copy of H is an isomorphic multigraph where the vertices and edges are relabelled in an increasing way (hence there is only one copy where the vertex and edge labels are consecutive integers starting at 1).

Lemma: Let G be a multigraph built from two copies of H sharing at least one vertex, then the density of G is greater than the density of H,

$$\frac{m(H)}{n(H)} > \frac{m(G)}{n(G)}.$$

Proof: Let J denote the largest common subgraph of the two copies of H in G. Then the number of vertices and edges in G are,

$$n(G) = 2n(H) - n(J), \quad m(G) = 2m(H) - m(J).$$

Since H is strictly balanced, the density of J is smaller than the density of H, so,

$$\frac{m(J)}{n(J)} > \frac{m(H)}{n(H)} \text{ which implies } \frac{m(G)}{n(G)} > \frac{m(H)}{n(H)}.$$

A patchwork is a set of copies of H, that might share vertices and edges (however, this is not a multiset, so two elements of a patchwork cannot be identical). The number of distinct vertices of a patchwork P is denoted by n(P), its number of distinct edges by m(P), and the number of multigraphs in P is denoted by |P|. The density of a nonempty patchwork is then m(P)/n(P). The generating function of patchworks is defined as,

$$P(z, w, u) = \sum_{\text{patchwork } P} u^{|P|} \frac{w^{m(P)}}{2^{m(P)} m(P)!} \frac{z^{n}(P)}{n(P)!}.$$

Lemma: The set of patchworks P ? that are either empty, or of excess m(H)/n(H), has generating function,

$$P^{*}(z,w,u) = \exp\left(u \frac{w^{m(H)}}{2^{m(H)} m(H)!} \frac{z^{n}(H)}{n(H)!} \right).$$

The density of all other patchworks is greater than the density of H.

Proof: A patchwork P is either a set of isolated copies of H, or contains at least two copies of H sharing at least a vertex. In the first case, P is either empty, or its density is equal to the density of H. In the second case, as a consequence of Lemma, the density of P is greater than the density of H. The generating function of a set of isolated copies of H is,

$$\exp\left(u \frac{w^{m(H)}}{2^{m(H)} m(H)!} \frac{z^{n(H)}}{n(H)!} \right).$$

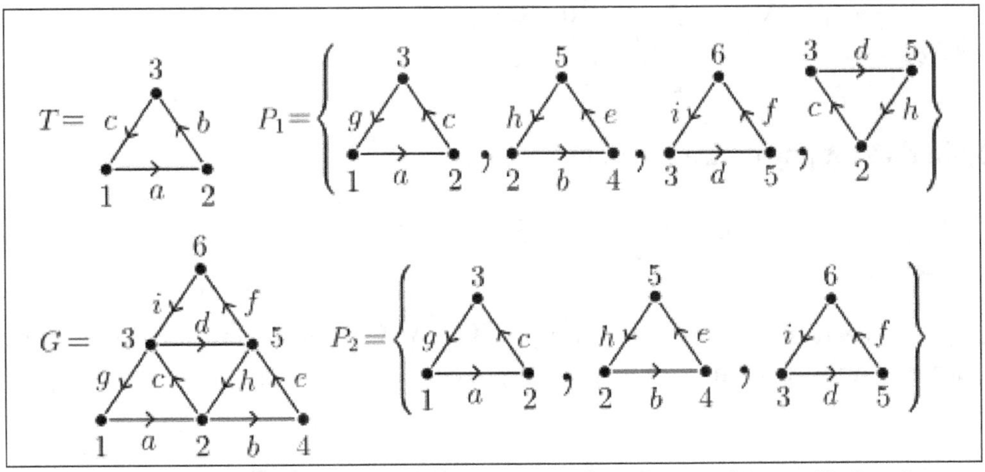

The multigraph H is here denoted by T. Two patchworks P_1 and P_2 are displayed. They both correspond to the same multigraph G.

Theorem: The generating function of multigraphs where a variable u marks the number of occurrences of subgraphs copies of H is,

$$MG(z, w, u) = \sum_{m \geq 0} (2m)! \left[x^{2m} \right] P\left(ze^{x}, w, u-1\right) e^{z \exp(x)} \frac{w^{m}}{2^{m} m!}.$$

Proof: The generating function of multigraphs where each occurrence of the subgraph H is either marked with the variable u, or left unmarked, is MG(z, w, u + 1). In such a multigraph G, by construction the set of marked subgraphs form a patchwork P. If we cut each edge that is not in P into two labelled half-edges, we obtain a representation of the multigraph G as:

- A patchwork P, where each vertex comes with a set of half-edges,
- A set of vertices (the vertices of G that do not belong to P), each attached to a set of half-edges.

The total number of half-edges must be even. Denoting this number by m, and using the variable x to mark the half-edges, we obtain,

$$MG(z,\ w,\ u+1) = \sum_{m\geq 0}(2m)!\left[x^{2m}\right]P\left(ze^{x},w,u\right)e^{z\exp(x)}\frac{w^{m}}{2^{m}m!}.$$

Lemma: The number of multigraphs with n vertices, m edges, and that have no subgraph that is a copy of H is,

$$n!2^{m}m!\left[z^{n}w^{m}\right]MG(z,w,0) = n^{2m}\frac{n!}{n^{n}}\frac{m!}{m^{m}}\sqrt{2m}\left[z^{n}x^{2m}\right]\frac{1}{\sqrt{2\pi}}\int_{-\infty}^{+\infty}P\left(nz,\frac{2m}{n^{2}}\frac{x^{2}}{t^{2}}-1\right)e^{nz}e^{2mx}t^{2m}e^{-mt^{2}}dt.$$

Proof: Applying the classic identity,

$$\frac{1}{\sqrt{2\pi}}\int_{-\infty}^{+\infty}t^{q}e^{-t^{2}/2}dt = \begin{cases}0 & \text{if } q \text{ is odd},\\ \dfrac{(2m)!}{2^{m}m!} & \text{if } q = 2m\end{cases}$$

we obtain that for any entire function f, we have,

$$\sum_{m\geq 0}(2m)!\left[x^{2m}\right]f(x)\frac{w^{m}}{2^{m}m!} = \frac{1}{\sqrt{2\pi}}\int_{-\infty}^{+\infty}f\left(\sqrt{wt}\right)e^{-t^{2}/2}dt.$$

We apply this relation to the expression of $MG(z,w,0)$, derived in Theorem. The number of multigraphs with n vertices, m edges, and without any subgraph copy of H is then,

$$n!2^{m}m!\left[z^{n}w^{m}\right]MG(z,w,0) = n!2^{m}m!\left[z^{n}w^{m}\right]\frac{1}{\sqrt{2\pi}}\int_{-\infty}^{+\infty}P\left(ze^{\sqrt{wt}},w,-1\right)e^{z\exp\left(\sqrt{wt}\right)}e^{-t^{2}/2}dt.$$

In order to apply the Laplace method and the large powers theorem with saddle-point 1, we transform this expression and apply successively the changes of variables,

$$z \to ne^{-\sqrt{wt}}z,\quad w \to \left(\frac{2m}{n}\frac{x}{t}\right)^{2},\quad t \to \sqrt{2mt},$$

The expression becomes,

$$n^{2m}\frac{n!}{n^n}\frac{m!}{m^m}\sqrt{2m}\left[z^n x^{2m}\right]\frac{1}{\sqrt{2\pi}}\int_{-\infty}^{+\infty}P\left(nz,\frac{2m}{n^2}\frac{x^2}{t^2},-1\right)e^{nz}e^{2mx}t^{2m}e^{-mt^2}dt.$$

Theorem: Set $\alpha = 2-\frac{n(H)}{m(H)}$, and consider integers n and m such that n/m^{α} has a positive limit c. Then the limit probability for a random multigraph with n vertices and m edges to contain no copy of H as a subgraph is,

$$\exp\left(-\frac{(2c)^{m(H)}}{m(H)!n(H)!}\right).$$

Proof: For simplicity, in this notes, we will assume that the generating function of patchworks satisfy the conditions of the Laplace method and the large powers theorems. Observe that when applying those techniques with saddle-points at 1, we have,

$$\int_c A(x)e^{en\phi(x)} \sim A(O)\int_c e^{n\phi(x)}, \text{ and } \left[z^N\right]A(z)B(z)^n \sim A(1)\left[z^N\right]B(z)^n.$$

Since we chose our changes of variables to ensure that the saddle-points are located at 1, we then obtain,

$$n!2^m m!\left[z^n w^m\right]MG(z,w,O) \sim n^{2m}P\left(n,\frac{2m}{n^2}-1\right),$$

where n^{2m} is the total number of multigraphs with n vertices and m edges. Therefore, the probability for a random multigraph with n vertices and m edges to contain no subgraph that is a copy of H has the same limit as $P\left(n,\frac{2m}{n^2},-1\right)$.

Since m is negligible compared to n^2, the dominant contribution comes from patchworks with a small density, so we consider only the contribution of patchworks of density smaller or equal to the density of H. According to Lemma, it is equal to,

$$\exp\left(-\frac{(2m/n^2)^{m(H)}}{m(H)!}\frac{n^{n(H)}}{n(H)!}\right).$$

It is then natural to consider m of the form cn^{α}, and we obtain,

$$\exp\left(-\frac{(2c)^{m(H)}}{m(H)!n(H)!}n^{(\alpha-2)m(H)+n(H)}\right).$$

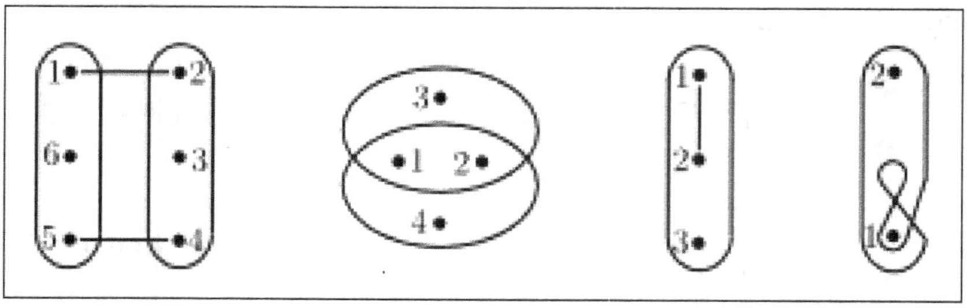

Four examples of hypergraphs.

If $\alpha < 2 - \frac{n(H)}{m(H)}$, then this exponential tends to 1, and almost no random multigraph with n vertices and m edges contains H as a subgraph. On the other hand, if $\alpha > 2 - \frac{n(H)}{m(H)}$, then this exponential tends to 0, and almost all multigraphs contain H as a subgraph. The value of interest is thus $\alpha = 2 - \frac{n(H)}{m(H)}$.. In this case, the limit probability for a random multigraph to not contain any subgraph copy of H is,

$$\exp\left(-\frac{(2c)^{m(H)}}{m(H)!n(H)!}\right).$$

Each graph with m edges corresponds to exactly $2^m m!$ multigraphs (the number of possible edge orientations and labelling). Conversely, any set F of multigraphs each with m edges, stable by edge relabelling and change of orientation, and that contain neither loops nor multiple edges, can be reduced to a set of $|F|/(2^m m!)$ graphs.

Hypergraphs and Inhomogeneous Multigraphs

Hypergraphs are a generalization of graphs, where each hyperedge can contain 2 or more vertices. They are used to represent databases: each vertex represents an object, and each hyperedge an attribute. Most of the work on hypergraphs focuses on the uniform case, where all hyperedges contain the same number of vertices. Using analytic combinatorics, de Panafieu adopted a more general setting, allowing any size of hyperedge in a given set D, and generalized to hypergraphs.

The inhomogeneous graph model is also known as the stochastic graph model, and is related to the Ising model. In this model, each vertex has a color, taken in a finite set, and only some colors are allowed to be linked by an edge. Those rules are encoded into a $\{0, 1\}$ symmetric matrix R. Properly k-colored graphs are a particular case, where each color can be linked to any different color. The matrix R is then the $k \times k$ matrix with 0 on the diagonal and 1's everywhere else. Any other Constraint Satisfaction Problem (CSP) where the constraints contain only two.

INFINITARY COMBINATORICS

In mathematics, infinitary combinatorics, or combinatorial set theory, is an extension of ideas in combinatorics to infinite sets. Some of the things studied include continuous graphs and trees, extensions of Ramsey's theorem, and Martin's axiom. Recent developments concern combinatorics of the continuum and combinatorics on successors of singular cardinals.

Ramsey Theory for Infinite Sets

Write κ, λ for ordinals, m for a cardinal number and n for a natural number. Erdős & Rado introduced the notation,

$$\kappa \to (\lambda)^{n}_{m}$$

as a shorthand way of saying that every partition of the set $[\kappa]^n$ of n-element subsets of into m pieces has a homogeneous set of order type λ. A homogeneous set is in this case a subset of κ such that every n-element subset is in the same element of the partition. When m is 2 it is often omitted.

Assuming the axiom of choice, there are no ordinals κ with $\kappa \to (\omega)^\omega$, so n is usually taken to be finite. An extension where n is almost allowed to be infinite is the notation

$$\kappa \to (\lambda)^{<\omega}_{m}$$

which is a shorthand way of saying that every partition of the set of finite subsets of κ into m pieces has a subset of order type λ such that for any finite n, all subsets of size n are in the same element of the partition. When m is 2 it is often omitted.

Another variation is the notation,

$$\kappa \to (\lambda, \mu)^{n}$$

which is a shorthand way of saying that every coloring of the set $[\kappa]^n$ of n-element subsets of κ with 2 colors has a subset of order type λ such that all elements of $[\lambda]^n$ have the first color, or a subset of order type μ such that all elements of $[\mu]^n$ have the second color.

Some properties of this include:

- $\aleph_0 \to (\aleph_0)^{n}_{k}$ for all finite n and k (Ramsey's theorem).
- $\beth_n^+ \to (\aleph_1)^{n+1}_{\aleph_0}$ (Erdős–Rado theorem.)
- $2^\kappa \not\to (\kappa^+)^2$ (Sierpiński theorem).

- $2^{\kappa} \nrightarrow (3)^2_{\kappa}$
- $\kappa \rightarrow (\kappa, \aleph_0)^2$ (Erdős–Dushnik–Miller theorem).

In choiceless universes, partition properties with infinite exponents may hold, and some of them are obtained as consequences of the axiom of determinacy (AD). For example, Donald A. Martin proved that AD implies,

$$\aleph_1 \rightarrow (\aleph_1)^{\aleph_1}_2$$

Large Cardinals

Several large cardinal properties can be defined using this notation. In particular:

- Weakly compact cardinals κ are those that satisfy $\kappa \rightarrow (\kappa)^2$.
- α-Erdős cardinals κ are the smallest that satisfy $\kappa \rightarrow (\alpha)^{<\omega}$.
- Ramsey cardinals κ are those that satisfy $\kappa \rightarrow (\kappa)^{<\omega}$.

ARITHMETIC COMBINATORICS

In mathematics, arithmetic combinatorics is a field in the intersection of number theory, combinatorics, ergodic theory and harmonic analysis.

Szemerédi's Theorem

Szemerédi's theorem is a result in arithmetic combinatorics concerning arithmetic progressions in subsets of the integers. In 1936, Erdős and Turán conjectured that every set of integers A with positive natural density contains a k term arithmetic progression for every k. This conjecture, which became Szemerédi's theorem, generalizes the statement of van der Waerden's theorem.

Green–Tao Theorem and Extensions

The Green–Tao theorem, proved by Ben Green and Terence Tao in 2004, states that the sequence of prime numbers contains arbitrarily long arithmetic progressions. In other words there exist arithmetic progressions of primes, with k terms, where k can be any natural number. The proof is an extension of Szemerédi's theorem.

In 2006, Terence Tao and Tamar Ziegler extended the result to cover polynomial progressions. More precisely, given any integer-valued polynomials $P_1, \ldots, P_k$ in one unknown m all with constant term 0, there are infinitely many integers x, m such that $x + P_1(m), \ldots, x + P_k(m)$ are simultaneously prime. The special case when the polynomials are m, 2m,..., km implies the previous result that there are length k arithmetic progressions of primes.

Example:

If A is a set of N integers, how large or small can the sumset,

$$A + A := \{x + y : x, y \in A\},$$

the difference set,

$$A - A := \{x - y : x, y \in A\},$$

and the product set,

$$A \cdot A := \{xy : x, y \in A\}$$

be, and how are the sizes of these sets related?

References

- De Longueville, Mark (2004), "25 years proof of the Kneser conjecture - The advent of topological combinatorics", EMS Newsletter, Southampton, Hampshire: European Mathematical Society, pp. 16–19, retrieved 2008-07-29
- De Panafieu. Phase transition of random non-uniform hypergraphs. Journal of Discrete Algorithms, 31 (0):26 – 39, 2015b
- Łaba, Izabella (2008). "From harmonic analysis to arithmetic combinatorics". Bull. Amer. Math. Soc. 45 (01): 77–115. doi:10.1090/S0273-0979-07-01189-5
- Flajolet, B. Salvy, and G. Schaeffer. Airy phenomena and analytic combinatorics of connected graphs. Electronic Journal of Combinatorics, 11(1), 2004
- Tao, Terence; Vu, Van H. (2006). Additive combinatorics. Cambridge Studies in Advanced Mathematics. 105. Cambridge: Cambridge University Press. ISBN 0-521-85386-9. MR 2289012. Zbl 1127.11002

PERMISSIONS

All chapters in this book are published with permission under the Creative Commons Attribution Share Alike License or equivalent. Every chapter published in this book has been scrutinized by our experts. Their significance has been extensively debated. The topics covered herein carry significant information for a comprehensive understanding. They may even be implemented as practical applications or may be referred to as a beginning point for further studies.

We would like to thank the editorial team for lending their expertise to make the book truly unique. They have played a crucial role in the development of this book. Without their invaluable contributions this book wouldn't have been possible. They have made vital efforts to compile up to date information on the varied aspects of this subject to make this book a valuable addition to the collection of many professionals and students.

This book was conceptualized with the vision of imparting up-to-date and integrated information in this field. To ensure the same, a matchless editorial board was set up. Every individual on the board went through rigorous rounds of assessment to prove their worth. After which they invested a large part of their time researching and compiling the most relevant data for our readers.

The editorial board has been involved in producing this book since its inception. They have spent rigorous hours researching and exploring the diverse topics which have resulted in the successful publishing of this book. They have passed on their knowledge of decades through this book. To expedite this challenging task, the publisher supported the team at every step. A small team of assistant editors was also appointed to further simplify the editing procedure and attain best results for the readers.

Apart from the editorial board, the designing team has also invested a significant amount of their time in understanding the subject and creating the most relevant covers. They scrutinized every image to scout for the most suitable representation of the subject and create an appropriate cover for the book.

The publishing team has been an ardent support to the editorial, designing and production team. Their endless efforts to recruit the best for this project, has resulted in the accomplishment of this book. They are a veteran in the field of academics and their pool of knowledge is as vast as their experience in printing. Their expertise and guidance has proved useful at every step. Their uncompromising quality standards have made this book an exceptional effort. Their encouragement from time to time has been an inspiration for everyone.

The publisher and the editorial board hope that this book will prove to be a valuable piece of knowledge for students, practitioners and scholars across the globe.

INDEX

www.ingramcontent.com/pod-product-compliance
Lightning Source LLC
LaVergne TN
LVHW080627210926
848485LV00003B/222
9781639892150